Smooth Sailing

AF695947

Popular Culture

Editor-in-Chief

Marcel Danesi (*University of Toronto*)

Associate Editors

Prisca Augustyn (*Florida Atlantic University*)
Lorenzo Cantoni (*USI—Università della Svizzera italiana*)
Paul Cobley (*Middlesex University*)
Ivan Kalmar (*University of Toronto*)
Annamaria Lorusso (*University of Bologna*)
Frank Nuessel (*University of Louisville*)
Jamin Pelkey (*Ryerson University*)
Susan Petrilli (*University of Bari*)
Deborah Smith-Shank (*Ohio State University*)
Hongbing Yu (*Ryerson University* and *Nanjing Normal University*)

Volumes published in this Brill Research Perspectives title are listed at *brill.com/rppcs*

Smooth Sailing

An Ethnographic and Socio-semiotic Analysis of Tourism and Ocean Cruising

By

Arthur Asa Berger

BRILL

LEIDEN | BOSTON

Library of Congress Control Number: 2022906579

Typeface for the Latin, Greek, and Cyrillic scripts: "Brill". See and download: brill.com/brill-typeface.

ISSN 2589-4420
ISBN 978-90-04-50627-5 (paperback)
ISBN 978-90-04-50626-8 (e-book)

Copyright 2022 by Arthur Asa Berger. Published by Koninklijke Brill NV, Leiden, The Netherlands.
Koninklijke Brill NV incorporates the imprints Brill, Brill Nijhoff, Brill Hotei, Brill Schöningh, Brill Fink, Brill mentis, Vandenhoeck & Ruprecht, Böhlau and V&R unipress.
Koninklijke Brill NV reserves the right to protect this publication against unauthorized use. Requests for re-use and/or translations must be addressed to Koninklijke Brill NV via brill.com or copyright.com.

This book is printed on acid-free paper and produced in a sustainable manner.

Contents

Acknowledgments

I would like to thank my editor, Jennifer Obdam, for her encouragement, fortitude, and help with the book and the production staff at Brill for their efforts. I also appreciate the suggestions made by two (unknown to me) reviewers of the book who offered incredibly detailed critiques of the book. Each of their reviews was five pages single-spaced, long, and one of the reviewers inserted notes, criticism, and suggestions on many pages of the manuscript draft. I owe a debt of gratitude to all the journalists, writers, and scholars whose ideas and writings I used in the book and to the many people I met in the course of my travels and cruises who provided me with their personal insights into tourism and cruising. Although publishing this book wasn't always smooth sailing, we were eventually able to do it which, in the final analysis, is what counts.

Illustrations

Smooth Sailing

An Ethnographic and Socio-Semiotic Analysis of Tourism and Ocean Cruising

Arthur Asa Berger
Broadcasting and Electronic Communication Arts,
San Francisco State University, California, USA
arthurasaberger@gmail.com

Abstract

Tourism and ocean cruising, which is an increasingly important kind of tourism and travel, can be considered forms of popular culture. Tourism is now the largest industry in the world and every year more than a billion people undertake travel to foreign lands. In addition, there is an enormous amount of travel within the borders of many countries. This study deals with topics such as what is it that motivates people to travel, what psychological benefits they get from travel, whether there is a difference between being a traveler and a tourist, what theories tourism scholars offer to explain the different kinds of tourists and why ocean cruising has become so popular? The preface has material on the devastating impact of the Covid-19 virus on the cruise industry and tourism, in general. These are some of the topics that *Smooth Sailing* deals with in an ethnographically informed socio-semiotic analysis of tourism and ocean cruising.

Keywords

tourism – travel – ocean cruising – psychoanalysis – ethnography – Covid-19 virus – myth – carnival – consumer cultures – quantum experience – semiotics

Preface on the Covid-19 Virus

The Covid-19 virus manifested itself in the United States in winter 2020 and devastated the tourism industry, the cruise industry, the restaurant industry, and many aspects of everyday life. In the United States, more than nine hundred thousand people died from becoming infected with the virus. Because of the virus, people stayed in their homes and stopped flying, stopped traveling,

© ARTHUR ASA BERGER, 2022 | DOI:10.1163/9789004506268_002

stopped going to restaurants, and stopped cruising. The news about the spread of the virus among passengers on cruise ships was terrible for the cruise industry. Some cruise lines, it should be said, kept sailing weeks after the coronavirus manifested itself and probably played a role in spreading it. It seems evident that these lines put profits over their ethical responsibilities, for which they paid a heavy price. The cruise industry faced economic challenges that were existential in nature.

If you think about it, a typical large cruise ship, with three or four thousand passengers and a thousand or more crew members, becomes very dangerous for everyone if just one person onboard has the virus. An article by Lawrence C. Strauss, "If Cruise Lines Survive, Can They Again Thrive?" tells the story. The subhead reads: "No-sail orders are draining away cash. Customers are spooked by images of floating 'Petri dishes.' Carnival, Royal Caribbean, and Norwegian are beaten down but not out." The article deals with the problems the big cruise lines faced in surviving in a period when the tourism industry, along with just about everything else, suddenly closed down.

An ocean liner can easily cost 500 million dollars and the largest ones cost more than a billion dollars. According to CruiseIndustryNews.com, there are 404 cruise ships (2019 figures) with more than 600,000 berths in an industry that served almost 28 million passengers. Then, suddenly, in March and April, the passengers all disappeared, leaving the cruise lines with payments for the ships and for the crew members and no income to speak of. The big lines survived by borrowing money and cutting expenses.

The Carnival cruise corporation has six new cruise ships being built, to add to the problems they have in filling the cabins in the ships they already have on the various lines Carnival owns, such as Princess and Holland America. And other cruise lines have just taken delivery of cruise ships and have other ships being built. It will be an interesting study in marketing and group psychology to see whether the cruise lines can find ways to attract cruise takers and start prospering again the way they did before the pandemic. It looks like they will.

A report on CNN Travel gives us the following insight:

> "The industry is going to return and is going to be stronger than ever," says Drew Daly, the senior vice president of *Dream Vacations*, a subsidiary of *World Travel Holdings*, the United States' largest cruise agency. "We are seeing people booking in 2021," says Daly. "Based on the same time last year, we're seeing greater growth." And analysts at UBS, a global financial advising firm, say bookings for 2021 cruises are up 9% over a recent 30-day period when compared with the same time last year. Both Daly and UBS note that their numbers include rebookings of coronavirus-related

> cancellations, but the website *Cruisecompete.com* told the *Los Angeles Times* that new bookings were up, too.

It may take some time to attract people who are scared of being on a boat with possible virus carriers, even if all the passengers have to show proof of having been vaccinated, but it is likely, as the bookings for 2021 and 2022 indicate, many people have overcome their fears of becoming infected by the virus and are taking cruises. Although no cruise ships were allowed to leave from American ports, many lines offered cruises from foreign ports in 2021 and cruises started leaving from American ports in the early summer of 2021.

The cruise lines are now using technology for contact-tracing passengers, even though most lines require vaccinations to take a cruise. As John McCormick explains in his article "Cruise Lines Turn to Contact-Tracing Tech" in the June 28, 2021 edition of *The Wall Street Journal* (page B5):

> **Royal Caribbean Group**, MSC Cruises and **Virgin Voyages** are among the cruise companies looking to phone apps, wearable devices, artificial intelligence and other technologies to keep passengers distances, which lessens the chance of airborne transmission of the virus, and to provide contact tracing if anyone does get sick.... Royal Caribbean ... recently developed a wrist device, the "Tracelet," to help with contact-tracing. Embedded with a Bluetooth-enabled chip, the Tracelet can record when a guest is within 6 feet for 15 minutes of other individuals on the ship, the company said. Royal Caribbean website states that all passengers and crew need to wear a Tracelet.

This article suggests that the cruise lines are worried about passengers getting sick, even if all of them (in many lines) have been vaccinated. Many lines are now offering cruises from Seattle to Alaska starting in July and it looks like the cruise industry will return to normalcy in the months to come.

It is ironic that this book is titled "Smooth Sailing," for that is just the opposite of what the tourism industry and the ocean cruising segment of that industry faced from 2020 to 2021. It is useful to assume that these years were rough waters and that as of summer, 2021 the tourism industry is recovering and the ocean cruising industry has re-entered smooth waters.

> All journeys have secret destinations
> of which the traveler is unaware.
>
> MARTIN BUBER

1 Introduction

At this point, let me say something about my style of writing. I offer a page of quotations at the beginning of each Part to give my readers something to think about relative to tourism and cruising. I also make use of quotations from numerous sources and use these quotations as a form of evidence that is meant to support my interpretations and assertions. I write in what I would describe as a conversational style that is accessible to most readers. I believe it is important to make books as easy to read as possible. While the book is not written in a "scholarly" style, in the narrowest sense of the term, it uses material from many theorists, thinkers, scholars, writers, and experts on various aspects of the tourism and cruising industries and the social sciences. You will find quotations from Sigmund Freud, Ferdinand de Saussure, Mikhail Bakhtin, Roland Barthes, Stanley Plog, Dean MacCannell, and many other important thinkers in the book. And now, since reading a book can be seen as a kind of journey, let us see where the trip I am offering everyone takes us.

Smooth Sailing is based on my travels with my wife, my writings, and my research on tourism and cruising over many years. It discusses topics such as why people travel and become tourists, what kinds of people travel and where they go, and what gratifications travel and ocean cruising offer them. The book is ethnographic in nature and combines research, psychoanalytic and socio-semiotic analysis, as well as personal experiences. It is a primer—an introduction to the subject that offers readers insights and information about the tourism industry and the ocean cruising sector of that industry. I have taken minor liberties with the paragraphing in many of the quotations to make the book more readable and I've added a number of my drawings and photographs to make the book more visually attractive.

Martin Buber's suggestion that there are "secret destinations" connected to our travels, of which we are unaware, is worth thinking about. We must take the notion of "secret destinations" in a broad sense. I think it fair to consider these secret destinations in terms other than places visited. These destinations involve our sense of ourselves and our relations with people in other lands. When we travel, we are seeking new experiences to satisfy our curiosity about how people in other countries live, to see historically important buildings and places, to gain a better understanding of history, to sample new cuisines, and many other reasons. Some of the things we search for are matters of which we are unaware, and are shaped by hidden imperatives residing in what Freud called the "unconscious." I will say more about this topic shortly. We travel to renew ourselves, to broaden our understanding of what life is all about, and to have chance experiences that are beneficial to us in unexpected ways. So travel is, at best, an adventure involving encounters with different people, different cultures, and different experiences that we could not have anticipated but which make travel so exciting.

When there are too many tourists visiting certain sites, problems arise—both for the tourists, who generally don't like being stuck in mobs of people, and for the inhabitants of the cities and towns where there is too much tourism. Too much of a good thing is, it turns out, bad for everyone involved. For example, Barcelona is trying to figure out how to limit the number of tourists who visit there and the same applies to many other places of interest to tourists. In *Travel Weekly, 7/10/2018* there is a discussion of overtourism that is of interest:

1.1 *Overtourism*

> OVERTOURISM ROUNDTABLE:
> At the World Travel and Tourism Council's Global Summit this year, a panel of eight travel industry executives discussed one of the most vexing problems they face: local resentment of and resistance to the overcrowding caused by the growing number of visitors to the world's most popular destinations.
>
> ARNIE WEISSMANN, EDITOR IN CHIEF, *TRAVEL WEEKLY*:
> "Coping with Success" identifies five symptoms of overtourism: alienated local residents, a degraded tourism experience, overloaded infrastructure, damage to nature and a threat to culture and heritage. Anything to add?
>
> CHRIS LEHANE, HEAD OF GLOBAL POLICY AND PUBLIC AFFAIRS FOR AIRBNB:
> What was missing, to me, was a definition of return on investment (ROI) from a community perspective. Tourism is, at some level, a natural resource. Is tourism driving broad-based benefits for the many and not just the few? And not just economic benefits, although I think that is at the front end of the conversation. I think one aspect of the overtourism question is a feeling, not a recognition or understanding, but the optical belief that you as a citizen are not benefitting from tourism in a meaningful way, yet your city's resources are being leveraged.
>
> *Travel Weekly*, 7/10, 2018er

Tourism, it seems, has impacted the lives of people who live in important tourism destinations, and not always in positive ways.

ILLUSTRATION 1 Tourists at Gaudi's cathedral La Sagrada Familia, Barcelona
PHOTO BY THE AUTHOR

There are, it turns out, many different kinds of tourists. Consider a number of the possibilities listed below (and there are others not listed):

1.2 *Kinds of Tourists*

Destination tourism	travel to culturally and historically important destinations
Disaster tourism	visit sites of disasters
Cultural tourism	go to concerts, plays, operas, museums
Religious tourism	visit cathedrals, mosques, synagogues
Food tourism	dine at great restaurants
Sports tourism	see important contests, sports events
Sexual tourism	find sexual partners
Family events tourism	attend weddings, family celebrations
Medical tourism	go for medical procedures
Adventure tourism	go to jungles and similar places
Sun tourism	escape cold and nasty weather

Sustainable tourism	prevent the degradation of sites
Educational tourism	study abroad
Events tourism	attend Mardi Gras and other celebrations

Many people combine several different kinds of tourism when they travel. The first part of this book deals with theoretical matters related to the nature of travel and tourism (are they different from one another?) and is drawn, here and there, from some of my previous writings. Tourism was an eight trillion-dollar industry in 2018 and when we are in a post-pandemic world, it will probably be even larger. In some cases, international travel can be to a country a few miles away but for many people, it involves traveling great distances in airplanes. For example, I live near San Francisco and it takes more than twelve hours to fly from San Francisco to Berlin.

As we might expect, many scholars in universities study tourism and offer countless theories and ideas about it in the many academic journals on the subject. I had the pleasure of spending two months as a visiting professor at the Hong Kong Polytechnic School of Hotel and Tourism Management many years ago. It is one of the leading schools of tourism and it was an amazing experience. At the school, there were around fifty professors who spent all their time (except when teaching) doing research, writing articles about different aspects of tourism, and advising graduate students at both the MA and Ph.D. levels.

Smooth Sailing will be of interest and use to anyone who likes to travel and wants to gain a better understanding of travel and tourism, and also information about an increasingly important aspect of tourism, ocean travel, and cruising. Tourism can be considered a form of mass culture or popular culture and ocean cruising, dealt with in the second part of this book, has become an increasingly important form of tourism.

2 Aspects of Travel and Tourism

> Self-contempt and a sense of fraudulence distinguish the attitude of contemporary self-conscious travelers. There is a touching desperation in the attempts of professional tourists, well-funded anthropologists, and recording travelers, to distinguish themselves from the traveling masses and run-of-the-mill adventurers. The most characteristic mark of the tourist is the wish to avoid tourists and places they congregate. But this is merely evidence of the fact that travel is no longer a means of achieving

> distinction. It is a way of achieving and realizing a norm, the common identity we all share—the identity of the stranger.
>
> ERIC J. LEED. *The Mind of the Traveler: From Gilgamesh to Global Tourism*, 1991: 286, 287

> I was possessed by a violent and insatiable desire to visit the places of God … as I had learned that eminent men such as Rabbi Isaac and Rabbi Solomon Levi had also been inflamed with the desire of accomplishing the holy journey. I, being urged on like them by a sort of divine instinct, did not lose sight of the execution of my project: I would not have suffered myself to be turned aside from it by any reason whatsoever. The desire to set out which had formed itself in my mind was so violent that it was impossible for me to remain in my own home, or to go about my accustomed business.
>
> SAMUEL JEMSEL. In Elkan Nathan Adler's *Jewish Travelers*, 1930

Many of my books on tourism have been about countries such as Vietnam, Thailand, Bali, India, and Japan. These books offer socio-semiotic analyses of the cultures in these countries and are not travel guides, though they do offer insights into aspects of these countries that any visitors would find interesting. So my travels, in many cases, have had a purpose other than purely sightseeing. Since my first trip abroad, at the age of twenty-two, one way or another, I've managed to travel to many places—just about all of the European countries, many Asian countries and North African countries, and some other countries where I've given lectures, such as Iran, Turkey, Lebanon, Israel, Brazil, Argentina, China, Ecuador, Vietnam, Thailand, and Morocco. I am now 88 so I've been traveling abroad for more than sixty years. I've visited countless countries over the years and in recent years my wife and I have taken two or three trips abroad annually. My wife was a professor, so we had plenty of free time to travel in summers and when I was on sabbaticals. I retired in 2003 and my wife retired in 2004 so we were free to travel wherever and whenever we wanted after 2004.

There are some people I've met who, somehow, are always on the move and have, it would seem, a compulsion to see as much of the world as possible. One of the ironic aspects of being a tourist, as the Leed quotation at the beginning of this Part points out, is that tourists seek to go to places of interest but prefer to go to places where there aren't other tourists. Or aren't too many other tourists. As I pointed out earlier, overtourism, as it is called by tourism professionals, is now a big problem. Big Sur and Venice along with many other tourist destinations now suffer from the blight. Some islands popular with cruise tourists are now trying to limit the number of cruise ships visiting the island to one per day. So there is a constant search by tourists to go to places before they have been "discovered." Some tourists seem driven when they travel and exhaust themselves trying to see every place of interest wherever they find themselves. As a result, they work harder on their vacations than they do in their ordinary lives. I will use the terms "travel" and "tourism" to represent traveling for pleasure and reserve "business travel" for when I'm writing about travel as part of work.

2.1 *Tourism as a Form of Consumption*

Many people now value personal experiences rather than consumer goods and have decided to spend their money on travel, in search of enriching experiences, as a means of satisfying their curiosity about how other people live and to satisfy some inner hunger to know more about the world. As tourists, we now "consume" other cultures and societies rather than just material goods. One of the most popular activities of tourists is shopping, so we can say there is a double valence to touristic consumption: tourists consume both places of interest to them and the objects of their desire. Because of cultural attitudes about work and vacations, world travel is typically more difficult for Americans than for some Europeans, who average many weeks of vacation each year and who have the time—assuming they have the financial resources—to go wherever they want. Many working Americans only average two (and sometimes three) weeks a year of vacations and thus find it much more difficult to travel to distant places. And many Americans don't take any vacations for one reason or another.

There are many gratifications that foreign tourism provides for people just as there are many different ways of traveling and countless places and countries to visit. How tourists make their decisions about such things is a matter of great interest to people in the travel business and to scholars who research tourism.

2.2 *Tourism as a Cultural Phenomenon*

This book examines tourism as a cultural phenomenon and attempts to understand why tourism has become the biggest industry in the world. Is foreign travel now a "rite of passage"? How does travel advertising represent other lands and give us a sense of the "exotic"? Is it correct to say that travel "broadens" people? And what about the different ways of traveling—airplanes, cars, trains, ocean liners? What are their appeals? Some of the topics I will deal with involve the search for "authenticity" in travel, the notion that tourism ultimately boils down to a desire to go to places that will make good photographs and videos, tourist meccas, and postmodern theories about travel. I will

ILLUSTRATION 2 Egyptian desert scene
PHOTO BY THE AUTHOR

also deal with travel and socio-economic class, travel and age, travelers as risk-takers, and with the impact of tourism on nature and Third World countries.

This book is informed by semiotics, the study of signs. A sign can be defined as anything that can be used to stand for something else. In essence, semiotics deals with how people find meaning in the world. Tourism can be seen as semiotic in nature: a means of finding (and often photographing and videotaping) the most important "signs" in the countries we visit such as the Taj Mahal, beautiful beaches in Bali, the Eiffel Tower, the White Desert in Egypt, and other historically important and aesthetically beautiful buildings and places. The photograph of the Egyptian desert, shown below, reflects the power that the landscape has on people. Being in that desert was, for me, an awesome and memorable experience.

3 Tourists and Travelers

> What gives value to travel is fear. It is the fact that, at a certain moment, when we are so far from our own country ... we are seized by a vague fear, and this instinctive desire to go back to the protection of old habits. This

> is the most obvious benefit of travel. At that moment we are feverish but also porous so that the slightest touch makes us quiver to the depths of our being.... This is why we should not say that we travel for pleasure. There is no pleasure in traveling, and I look upon it as an occasion for spiritual testing. Pleasure takes us away from ourselves the same way that distraction, as in Pascal's use of the world, takes us away from God. Travel, which is like a greater and grave science, brings us back to ourselves. (pp. 13, 14)
>
> ALBERT CAMUS. *Notebooks, 1935–1942*

> World War Two marked the end of the great age of travel. And it is easy to see why. Not only did the war make travel difficult, but the world that emerged after it—a monoculture of mass consumerism, package-holidays, and the extraordinary imperialism of the American hamburger—left little room for the sensibility characteristic of so many of the writers of the twenties and thirties. The books of Robert Byron, say, or Waugh, Auden, and Priestly are virtually Edwardian in temperament: they hark back to an earlier time. They wrote, as Paul Fussell has pointed out in *Abroad*, to celebrate a "Golden Age": the corners of the world uncontaminated by the manifestations of modernity. "One travelled," Fussell says, "to discover the past."
>
> BILL BUFFORD, in *Granta: Travel Writing*, 1984

People have wandered from place to place on the earth for many different reasons. We know, from archaeological investigations, that there have been incredible migrations of our ancient ancestors over the millennia. And nowadays, people continue to travel. Why they travel and what it all means is the subject of the first part of this book.

3.1 *Travelers and Mythology*

It was Moses, who fled from Egypt and settled in Midian, who described himself then, as many travelers might describe themselves now when they travel to foreign countries, as "a stranger in a strange land." Indeed, the history of the Jewish people is that of wanderers—after having been thrown out of Spain and forced to leave many other countries since the Middle Ages. But Jews are not the only people who have been forced to wander around the world, and even today, there are incredible migrations of peoples escaping war or famine. There are millions of people who, like Moses, also find themselves as strangers in strange lands but they are not tourists but immigrants seeking to find a way to survive.

ILLUSTRATION 3
Roland Barthes
DRAWING BY THE AUTHOR

There is a connection, I believe, between tourism and myths. Students of mythology have suggested that there is often a hidden and unrecognized mythological root to many of the things we do, such as visiting sites with religious significance. Myths are narratives that, among other things, connect individuals with their cultures and with various activities that take place in their cultures. Mark Shorer, a distinguished literary critic, defined myths as "instruments by which we continually struggle to make our experience intelligible to ourselves. A myth is a large, controlling image that gives philosophical meaning to the facts of ordinary life." But myths can have other functions, according to semiotician Roland Barthes.

In *Mythologies*, the French semiotician Roland Barthes has several chapters on travel. The first, "The Writer on Holiday," deals with attitudes people have about writers on holiday. Barthes writes (1972: 29):

> Holidays are a recent social phenomenon, whose mythological development, incidentally, would be interesting to trace. At first a part of the school world, they have become, since the advent of holidays with pay, a part of the proletarian world, or at least the world of working people. To assert that this phenomenon can henceforth concern writers, that the specialists of the human soul are also subjected to the common status of contemporary labour, is a way of convincing our bourgeois readers that they are indeed in step with the times.

Ultimately, Barthes concludes, "the good-natured image of 'the writer on holiday' is therefore no more than one of those cunning mystifications which the

ILLUSTRATION 4 Japanese geishas
PHOTO BY THE AUTHOR

Establishment practices to better enslave its writers." The title of his book refers to his view that French culture is full of myths that mystify people and support the status quo.

Barthes wrote a semiotically informed "travel" book about Japan, *Empire of Signs*, in which he analyzed important signifiers of Japan such as tempura, Japanese chopsticks, bowing, Japanese bodies, and Pachinko. For Barthes, these topics were his way of capturing Japan. He explained his methods as follows (1982: 4):

> The author has never, in any sense, photographed Japan. Rather, he has done the opposite: Japan has starred him with any number of "flashes;" or, better still, Japan has afforded him a situation of writing.

What he means by this is that he is not attempting a historical analysis of Japan but a series of essays on the semiology of Japan: important Japanese signs that offer insights into Japanese culture and society. He explains that he has isolated (1982: 3) "a certain number of features (a term employed in linguistics) and out of these features deliberately form a system. It is this system which I shall call: Japan."

My book takes a cue from *Empire of Signs* and uses semiotics to deal with many aspects of what Barthes would call "features" of a semiotically rich topic: tourism and travel and a form of travel, ocean cruising.

3.2 *Travel and Myth*

In *Mythologies*, Barthes ties travel on ships to the mythic—namely Jonah's ark—and hints at an important element of travel, namely the desire of people from various social classes to maintain social distance from those not like themselves. I will return to this matter later in the book. Mircea Eliade, a scholar of religion who has written extensively on the subject of myths, describes myths as sacred histories and suggests in *The Sacred and The Profane* that they are "the paradigmatic model for all human activities" (1961: 97,98). He also has written, "The modern man who feels and claims that he is nonreligious still retains a large stock of camouflaged myths and degenerated rituals" (1961: 204, 205). He offers as examples, things like New Year's Eve parties, parties given when one is promoted, battles between monsters and heroes in films. Another myth scholar, Raphael Patai, writes in *Myth and Modern Man* (1972: 2):

> Myth … is a traditional religious charter, which operates by validating laws, customs, rites, institutions, and beliefs, or explaining socio-cultural situations and natural phenomena, and taking the form of stories, believed to be true, about divine beings and heroes … Myths are dramatic stories that form a sacred charter either authorizing the continuance of ancient institutions, customs, rites and beliefs in the area where they are current, or approving alterations.

Patai adds that myths play an important role in shaping social life and writes that (1972: 2) "myth not only validates or authorizes customs, rites, institutions, beliefs, and so forth, but frequently is directly responsible for creating them."

I have developed what I call a "myth model" which argues that myth not only informs many of our daily activities but also plays a role in other matters, even if we might not recognize it or be aware of it as doing such. Let me suggest that we can take a myth and find that myth informing:

- Psychoanalytic theory,
- Historical experience,
- Elite culture,

ILLUSTRATION 5 Roman mosaic of Odysseus chained to the ship's mast to prevent him from going to the Sirens. The House of Dionysus and Odysseus, Thugga (now Dougga), Tunisia. Roman civilisation, 2nd–3rd century AD Bardo National Museum, Tunisia
PHOTO BY THE AUTHOR

– Popular culture (and)
– Everyday life.

Here is an example of how myth pervades different aspects of culture as reflected in my "myth model."

The model will deal with the mythological hero, Odysseus—also known as Ulysses—as a paradigmatic figure, who can be seen as the UR traveler, the model for all travelers, even though most of us don't take such extended trips or encounter such dangers.

3.3 *The Myth Model and Travel/Tourism*

The myth model always begins with a myth and then investigates, as I explained earlier, how that myth has informed various aspects of psychoanalytic theory, history, elite culture, popular culture, and everyday life. The myth I have chosen for this analysis of travel and tourism is the myth of Odysseus/Ulysses, which is found in Homer's classic book, *The Odyssey*. It begins as follows:

> The hero of this tale which I beg the Muse to help me tell is that resourceful man who roamed the wide world after he had sacked the holy citadel of Troy. He saw many cities and learned their ways. He suffered many hardships on the high seas in his struggles to preserve his life and bring his comrades home.

THE MYTH

Odysseus goes to Troy, spends ten years there, fighting in the Trojan War, and returns on a voyage that takes another ten years, and is full of hazards and dangers. Homer's book, *The Odyssey*, can be looked upon as the UR travel book. The image, shown above, is from a mosaic about Odysseus.

PSYCHOANALYTIC MANIFESTATION OF THE MYTH

Wanderlust—the need some people have to travel, endlessly. Freudian psychoanalytic theorists have speculated that wanderlust is tied to the unconscious desire of the wanderer to find the "good" father. His wanderings represent his search, at the unconscious level, for this figure or, as others have suggested, the "good" mother.

HISTORICAL EVENTS AND THE MYTH

A Frenchman, Alexis de Tocqueville, travels to America and later writes a book about American culture and politics, *Democracy in America*. Countless other examples of books by travelers could be offered here.

ELITE CULTURE MANIFESTATIONS OF THE MYTH

Thomas Mann's "Death in Venice," Jonathan Swift's *Gulliver's Travels*, and many others. Travel writings by distinguished authors, novelists, and travel writers such as Gabriel Garcia Marquez, Jonathan Raban, and Salman Rushdie.

POPULAR CULTURE MANIFESTATIONS OF THE MYTH

Travel magazines, television travel shows, travel guidebooks about any city or country. Also, films that take place in beautiful locations that stir viewers to travel to them, such as "Lord of the Rings," which led to increased travel to New Zealand.

EVERYDAY LIFE MANIFESTATIONS OF THE MYTH

A person takes a cruise in the Mediterranean on Viking Ocean Cruises or some other line.

What this little exercise in mythic analysis suggests is that there may be mythological foundations that become removed from their sacred origins behind many of the things we do—one of which involves travel. That is, many of our activities are what Eliade has described above as "camouflaged" myths and rituals and one of those camouflaged myths and rituals involves travel. Note, also, that there were dangers that Odysseus faced in his travels—dangers that heightened the significance of his voyage just as interruptions, disruptions, and a certain element of anxiety about one's safety heighten the experience of travel for many contemporary tourists. Tourists are often swindled by taxi cab drivers, shop owners, and others in the course of their travels, face delays in their flights and face other anxiety-provoking problems.

3.4 *What Is Travel?*

For our purposes, travel and tourism involve people going somewhere from somewhere else and then returning. When we commute from our houses to our jobs we are literally speaking traveling, but not in the sense that I am using the term in this book, which involves a break in our everyday routines and movement to near or distant places, for short periods, and then a return to one's starting point. Tourism, as I interpret the term, is a leisure time experience and its goal is pleasure, entertainment, adventure, renewal, and education.

Many people do a great deal of traveling who don't fit under my definition. There is, for example, business travel—which involves moving from place to place, but essentially to conduct business. Many members of the United States armed forces do a great deal of traveling—to fight wars, to staff military outposts all over the world. But this can be seen as a form of business travel. But sometimes businesspersons and soldiers have time to be tourists and visit cathedrals, temples, museums, and sites, which means they can mix tourism with business travel.

Travel or what we might call "leisure-time touristic travel" is associated with attributes like an adventurous personality, independence, and curiosity. The Camus quote alludes to the fact that travelers frequently must deal with delays of one sort or another, must make changes in their plans, and sometimes they may even face danger—all of which heighten their feelings of adequacy and of being alive. There is generally an element of physical discomfort and sometimes even "pain" connected with international travel; travelers become exhausted from long flights necessary to travel great distances. Travel is hard work, as every traveler knows—even if you are on a cruise ship and don't have to worry about finding a place to stay and a good restaurant. You have to get to the cruise ship and that often involves long flights.

3.5 *Are Tourists Travelers?*

Many people think of themselves as "travelers" and not "tourists," perhaps because the connotations of "tourist" are not positive. Travel, they believe, suggests effort and adventure while tourism suggests ease. But what about adventure tours, which can be quite strenuous? Even many ordinary tours, because of their rapid pace and the sightseeing involved, are also fairly strenuous. So is there a difference between being a traveler and a tourist? Not as I use the terms. People travel to foreign lands and some of these travelers may take organized tours of one sort or another. Those who travel on business I describe as business travelers—they travel because they must and their goal is not seeing the sights where they go but taking care of business matters. But in their free time, they may become tourists.

The root of the term, "tourist" is the Greek word *tornos*, which is a tool for making a circle. A "tour," as it is conventionally used in the travel industry, is a form of group travel with the following characteristics: the route and the places one visits are very specific; one moves from point to point or city to city rather quickly ("If it's Wednesday, this must be Lisbon"); there is a certain element of regimentation involved ("Will everyone please get on the bus!"), and eventually, one returns to where one started. But there is no universally accepted definition of tourism and different scholars define it in different ways. The World Tourism Organization defines tourism as follows:

> It comprises the activities of persons traveling to and staying in places outside their usual environment for not more than one consecutive year for leisure, business, and other purposes not related to the exercise of an activity remunerated from within the place visited.
>
> http://www.world-tourism.org/statistics/tsa_project/TSA%20depth/chapters/ch3-1.htm

I find this definition too broad and I will offer some ideas that will provide more insights into what tourism is and isn't. For many people, as pointed out earlier, the term tourism has negative connotations, because of the supposed superficiality and regimentation of some package tours and because of the connection of tourism with consumer culture.

In his seminal book, *The Tourist: A New Theory of the Leisure Class*, Dean MacCannell defines the tourist as the model for modern man and woman (1976: 1):

> Tourist is used to mean two things in this book. It designates actual tourists: sightseers, mainly middle-class, who are at this moment deployed through the entire world in search of experience. I want the book to serve as a sociological study of this group. But I should make it known that, from the beginning, I intended something more. The tourist is an actual person, or real people are actually tourists. At the same time, "the tourist" is one of the best models for modern-man-in-general. I am equally interested in "the tourist" in this second, metasociological sense of the term. Our first apprehension of modern civilization, it seems to me, emerges in the mind of the tourist.

For MacCannell then, tourists are mainly middle-class sightseers in quest of experience, who also serve as models or prototypes for modern man. This would suggest, if you push this notion to its logical conclusion, that to be a human, or at least, a modern man (and woman) is to be a tourist. Homo sapiens (man and woman, the thinker) has been replaced by homo tornos (man and woman, the traveler and tourist).

3.6 *Erik Cohen's Typology of Tourists*

Erik Cohen, an Israeli sociologist who has written many important articles and books on tourism, has an interesting typology of the different kinds of tourists which I describe as follows:

THE ORGANIZED MASS TOURIST:
Takes tours, is in a "bubble"
THE INDIVIDUAL MASS TOURIST:
Makes own arrangements
THE EXPLORER:
Seeks involvement with people in foreign lands
THE DRIFTER:
Tries to live like natives

Cohen wrote an article, "Toward a Sociology of International Tourism" (quoted in Kotler, et al., *Marketing for Hospitality and Tourism*, Second Edition) in which he explains what he means by organized mass tourists (1999: 662):

> The organized mass tourist is the least adventurous and remains largely confined to his "environmental bubble" throughout his trip. The guided tour, conducted in an air-conditioned bus, traveling at high speed through a streaming countryside, represents the prototype of the organized mass tourist. This tourist type buys a package tour as if it were just another commodity in the modern mass market. The itinerary for his trip is fixed in advance, and all his stops are well-prepared and guided: he makes almost no decisions for himself and stays almost exclusively in the microenvironment of his home country. Familiarity is at a maximum, novelty at a minimum.

This kind of tourism appeals to people who have anxieties about foreign travel and want to minimize all risks. I think it is a rather extreme characterization of these kinds of tourists because many people on organized tours still manage to spend time on their own away from the tour and interact with people they encounter in foreign lands.

3.7 *Are There Important Differences between Tourists and Travelers?*

There is the question of whether there are important differences between tourists, travelers, visitors, and excursionists. Visitors are generally defined as people who spend some time in a foreign country. But how long must a person stay in a distant place to stop being considered a visitor? And how do we—or can we—really differentiate visiting from traveling and touring? Excursionists are people who spend a short time at a site—such as those who take a day trip to a different city or visit a city or place of interest while on a cruise.

Travelers go places and then come back, but so do tourists. One thing we can say is that all tourists are travelers but not all travelers are tourists. I'm thinking of business travelers here, who I don't classify as tourists or people who visit foreign countries on their own and don't take tours. So, what is a tourist? What is tourism? International tourism is generally understood to have the following characteristics:

- It is *temporary*, for a relatively short period.
- It is *voluntary*, done by choice. (Some describe tourism as "voluntary suffering").
- It is done in *foreign* lands or places outside of one's usual environment.
- It is tied to *leisure* and *pleasure* and *consumer culture*.
- It is *discretionary*, done because the tourist desires to do so.
- It is done for *pleasure, entertainment, adventure*, and related considerations.
- It is *not involved with business* and earning money while abroad.
- It is based on a *round trip*, tied to a return to one's point of origin,
- It is made possible by *technological developments* in transportation and communication, among other things. (The jet airplane, the cruise ship, and the Internet play an important role in tourism now.)
- It is a *mass phenomenon* involving large numbers of people.

ILLUSTRATION 6
La Sagrada Familia by Antoni Gaudi, Barcelona
PHOTO BY THE AUTHOR

But what do we say about a business traveler in Dusseldorf who takes a couple of hours off to go to Cologne to look at the great cathedral there? Do we not have here a mixture of travel and tourism? Or what about a tourist who meets with someone, for a short time, about a business matter during a visit to a foreign land? Does a short period devoted to business make a tourist a traveler? If you live in San Diego and were to drive to Ensenada, Mexico for lunch and then come back, were you a tourist or a traveler or a visitor?

One of my friends recently wrote to me that he was going to Barcelona for a conference but hoped to take a few days more to visit the important sites there. He is, technically speaking, a business traveler, but for a few days, he will also be a tourist. He will most likely, among other things, visit the fantastic La Sagrada Familia cathedral by Antoni Gaudi in Barcelona, its most important tourist site. There are also walking tours of sites that are devoted to Gaudi's buildings (but you must take public transport to get to his cathedral).

It is obvious then, that it is frequently very difficult to differentiate between travel and tourism except, perhaps, in their most extreme forms. The differences between the two have historical roots. A hundred years ago, travel was more or less reserved for elites; when mass transportation developed—railroads, airplanes, cruise ships—travel then became available to the masses and this kind of travel, when it is linked with the phenomena described above, is now often described as tourism. Tourism, like beauty, ultimately is in the eye of the beholder and often in the mindset and ideological belief system of the person labeling some people as "tourists" and others as "travelers."

Some might say that a traveler is someone who goes places and a tourist is someone who sees places, but that distinction seems a bit forced because when travelers get to wherever they are going, they like to look around and see the sights, as well. And tourists have to travel to get where they want to go. So I consider the terms to be synonymous.

3.8 *Tourists as Continual Spenders*

I would suggest that tourism—now a global phenomenon—is the locus of a huge number of businesses connected to the movement of people from one place to another and with taking care of their many needs.

I am talking about everything from taking cabs and buses to and from airports, booking airplane travel, staying in hotels and inns, dining in restaurants, shopping (for souvenirs, gifts, and other items), visiting museums, going to concerts, operas, shows, and sporting events, finding sexual partners and taking day tours and excursions. In some cases, travelers go to hotels booked by their tours, and then their daily expenses are taken up by tour companies, with whom these travelers have booked their tours. They have paid, up-front, for their hotels, transportation, and many of their meals, or, in some cases, all of their meals. In other cases, independent travelers spend money on the go, so to speak, for these matters. We see, then, that travel, per se, is just the tip of a huge iceberg involving many different institutions and organizations, and companies connected to the tourist industry.

One way or another—in payments upfront to tour companies or cruise lines or in payments at each step of the process—money is changing hands. What tourists are buying, primarily, are experiences—ones that they believe will educate them in terms of the ways that other people live, elevate them spiritually, enhance their enjoyment of life and, in the broadest sense possible, entertain them. This is done by visiting, for varying lengths of time, other countries, to get a sense of how people in these places live, what they have done in the past (their architecture, monuments, and historically interesting buildings), what kinds of things they eat, and how they amuse themselves, among other things.

Paul Fussell explains this link between tourism and education in his book, *The Norton Book of Travel*. He writes, after discussing Freud's idea that travel involves an escape from the family and the father (1987: 13–14):

> But if travel offers the thrill of quasi-felonious escape, it also conveys the pleasure of learning new things, and as Aristotle observed over 2,300 years ago, not only philosophers but people in general like learning things, even if the learning comes disguised as "entertainment." It is as learners that explorers, tourists, and genuine travelers, otherwise so different in motives and behavior, come together. Explorers learn the contours of undiscovered shorelines and mountains, tourists learn exchange rates and where to go in Paris for the best hamburgers.... Travelers learn not just foreign customs and curious cuisines and unfamiliar beliefs and novel forms of government. They learn, if they are lucky, humility. Experiencing ... a world different from their own, they realize their provincialism and recognize their ignorance.

The ultimate lesson we learn from our travels abroad, Fussell suggests, is humility, as we discover that people in other countries may have beautiful cities full of wonderful buildings and great museums. People in other lands also may drink better coffee and have tastier food than many of us eat in America. One thing that people are searching for when they travel abroad is differences from their everyday lives because we only make sense of things by seeing how they differ from other things.

Ferdinand de Saussure, one of the founding fathers of semiotics, said that in language there are only differences and I'd say the same for tourism: tourists

ILLUSTRATION 7
Ferdinand de Saussure
DRAWING BY THE AUTHOR

are in search of differences from their everyday lives. If you go somewhere and it isn't different in interesting ways, why bother?

He also explained that concepts have no meaning in themselves but only obtain meaning by being different from their opposites. As he explained in his *Course in General Linguistics* (1966: 114, 117–118 [my italics]):

> Language is a system of interdependent terms in which the value of each term results solely from the simultaneous presence of the other. It is understood that … *concepts are purely differential and defined not by their positive content but negatively by their relations with the other terms of the system. Their most precise characteristic is in being what the others are not.* … Signs function, then, not through their intrinsic value but through their relative position.

The most important relationship for Saussure is opposition, "*being what the others are not.*" [My italics.] We make sense of things, in our everyday lives, by putting everything into a set of relationships and determining what something is by seeing what it isn't. The same applies to tourism; we travel to experience differences, ranging from countries that are mildly different from America (Canada) to countries that are very different from America (Thailand, Bali, India), which we define as "exotic." In a sense, we go traveling because we want to find out more about our own countries and in the case of the United States, we do this by seeing how other countries differ from it. We also travel for other reasons, such as to meet and interact with people in other lands and in so doing to discover our common humanity, to see beautiful sites, to visit great buildings and see works of art in great museums.

Travel is based upon a powerful desire to participate in history—to see where great events took place, to see where great writers and artists and scientists lived and to see how they lived, to experience the best that the world can offer us. So, we expect to get a great deal for our money, and many people feel that their travels and tours were worth the expense. Even people who have not been satisfied, for one reason or another, with their travel arrangements and with what happened on their travels and tours tend to convince themselves that their travels were worth the expense. Otherwise, they would have to admit to themselves that they made a mistake in booking a tour with this or that company and going to the country they visited rather than somewhere else. When you are a tourist, you are always plagued by a sense that you chose the wrong country to visit or the wrong time to visit it, or that you chose the wrong dish from the restaurant menu. That is, tourists always suffer from FOMA—the fear of missing out.

3.9 *Explorers, Settlers, Travelers, and Tourists*

Perhaps it is possible to distinguish between travelers and tourists differently—over periods of time. I offer here a short, hypothetical analysis of the relationship that might exist among all of the above, based on the American experience. Explorers can be understood as people who venture into lands where few others have gone, mapping the territory and interacting with native peoples—if there are any. Explorers do not know what they will find and often there is an element of adventure and danger connected to exploration. In the early days of America, explorers ventured into a great wilderness, what we called "the frontier."

Once the explorers came back and told others about their adventures, then others came who settled the frontier and brought "civilization" to it. Then came travel and tourism. Of course, some people visit so-called "unknown" lands that are already settled, lands that may have highly developed cultures and social institutions. The information explorers provide about these lands brings travelers and the information travelers provide often leads to large numbers of people visiting a given place—that is, to mass tourism. Tourism is a "mass" phenomenon in that huge numbers of people now travel and visit near or distant parts of their own countries or other lands. It is also very complex because it requires many institutions to sustain itself—what we might call a touristic infrastructure: travel agents, hotels and inns, museums, sports arenas, restaurants, entertainment sites, natural beauty sites, guides, transportation facilities, and so on. The world, it can be argued, has been turned into a gigantic collection of tourist sites, of varying natures, to suit the inclinations of different kinds of tourists.

3.10 *Stanley Plog on Different Kinds of Tourists*

One of the most famous scholarly articles ever written on tourism deals with the psyches of different kinds of tourists. The author, Stanley Plog, suggests there are five different kinds of tourists, based on their adventurousness or fearfulness, which he discusses in his article, "Why Destinations Rise and Fall in Popularity." I describe Plog's five types of tourists as follows:

ALLOCENTRIC
Tolerate considerable risk and are adventurous

NEAR ALLOCENTRIC
Tolerate some risk

MID-CENTRIC
Halfway between risk-taking and fearfulness

NEAR PSYCHOCENTRIC
Somewhat fearful

PSYCHOCENTRIC
Very fearful, do not tolerate risk.

Allocentric tourists discover new sites of interest and when they are fully developed and safe, psychocentric tourists will visit them. Visiting Europe, the United States, and many countries in Asia, such as Japan and Thailand is a psychocentric experience as there is relatively no risk involved in traveling there. When we travel, we often describe what we are doing as taking a trip.

Below I offer an itinerary for a fifteen-day trip to China that my wife and I took in September 2018. This trip was what Cohen would describe as organized mass tourism in which we were, more or less, in a tourist bubble. We spent a lot of time on tour buses and visiting factories where we had buying opportunities.

3.11 ***China Tour***

Thurs.	Sept. 20	Fly from SFO to Beijing
Fri.	Sept. 21	Arrive in Beijing
Sat.	Sept. 22	Free Day in Beijing
Sun.	Sept. 23	Great Wall of China trip
Mon.	Sept. 24	Beijing to Shanghai to Suzhou
Tues.	Sept. 25	Suzhou to Wuxi
Wed.	Sept. 26	Wuxi to Hangzhou
Thurs.	Sept. 27	Hangzhou to Shanghai
Fri.	Sept. 28	Shanghai
Sat.	Sept. 29	Shanghai to Yichang. Board River Cruise Ship.
Sun.	Sept. 30	Yangtze River Cruise
Mon.	Oct. 1	Yangtze River Cruise
Tues.	Oct. 2	Yangtze. River Cruise
Wed.	Oct. 3	Chongqing Visit
Thurs.	Oct. 4	Chongqing to San Francisco

ILLUSTRATION 8 Shanghai, China
PHOTO BY THE AUTHOR

This itinerary covered many of the more important cities and tourist sites in China, such as the Great Wall, Shanghai, and Suzhou, and had several stops at factories where we had "shopping opportunities" and many places where we had "photo opportunities," such as in the Bund in Shanghai and the Great Wall. Because the tour was so structured I had a sense that I was, in a curious way, a prisoner of the tourist guide, who controlled my life and everyone's life during the tour. But we did have some free days and they were very important psychologically.

3.12 *Tourists Take Trips*

Tourists take trips. That is, they go by themselves or take tours or cruises or "expeditions" to certain places—what I call "meccas" for specific periods, at the end of which they return home. People who are retired and who like to travel can take trips whenever they want, but those who aren't retired have to fit their trips into their vacation periods. Since Americans have shorter vacations than Europeans, the trips have to be planned carefully. The average worker in some Germany, for example, has six weeks of vacation compared to two or three weeks for the typical American worker. Here are some statistics of interest about workers in the United States (all figures are approximate, taken from various sources on the Internet):

- 16% of American workers say they are too busy to take off all the vacation days that are due to them.
- 20% of American workers never take vacations.
- 13 days of vacation is the average in the USA and 42 days in Europe.
- 33% of American workers check in with their bosses during vacations.
- 40% of American workers skip summer vacations because of increased workloads and tighter budgets.

An article by Hannah Sampson, "What does America have against vacations?" in *The Washington Post* (August 28, 2019) discusses the American aversion to vacations:

> According to the study from Center for Economic and Policy Research, the European Union requires member countries to grant workers at least 20 working days of paid vacation. But many nations go well above that number, and some offer a heap of paid holidays, to boot. France, for example, requires at least 30 paid workdays off, not including paid holidays, while the UK mandates 28, followed by Austria, Denmark, Finland, Norway, Spain and Sweden at 25....
>
> The United States, on the other hand—on the *very worst* hand—mandates no paid vacation or paid holidays. Zero days. It's the only country in the Organization for Economic Cooperation and Development, a group of 36 of the world's wealthiest nations, that doesn't require employers to give workers annual paid leave, according to the Center for Economic and Policy Research. That leaves 23 percent of Americans with no paid vacation and 22 percent without paid holidays.
>
> https://www.washingtonpost.com/travel/2019/08/28/what-does-america-have-against-vacation/, Accessed 6/29, 2021

We see, then, that Americans (and I will use this term in this book to mean people from the United States) are very hard-working people who don't, as a rule, have the amount of free time that workers in European countries have. On the other hand, Americans are relatively affluent and take a high percentage of all cruises and tours.

Though we don't have as much vacation time as workers in Europe, we do manage to do a good deal of traveling. This is because some jobs, such as teaching, provide a considerable amount of free time, and retired people often do a lot of traveling. And people with only two weeks of free time often find ways to squeeze in trips longer than two weeks and cruises of varying lengths. Many cruises are, for example, a week or ten days in length, which makes it possible, though not always easy, for Americans to take them.

There are, then, at any moment, huge numbers of people wandering around the world, functioning as tourists and travelers, of one sort or another. I am talking about people on vacations, students who take time off from their studies to travel, workers who take "sabbaticals" (without pay) to travel, students studying in foreign universities, elderly people taking tours, Australians with around-the-world air tickets (who often work in the UK for a year or two), and backpackers seeing the world on the cheap.

They all want to visit the interesting places of the world and often find themselves in what I call tourist "meccas," places of particular interest for one reason or another. For example, on a recent trip to Malaga, my wife and I went to Ronda, a tourist mecca that was mobbed with tourists who wanted to experience Ronda's remarkable beauty.

3.13 *Tourist Meccas*

Tourist meccas, as I will use the term, are "choice" places for tourists to visit, because of the beauty of the scenery, the richness of the culture, certain events that may be taking place (carnivals, sporting events), and other similar things. The term "mecca" comes from Mecca, the city in Saudi Arabia that millions of Muslims visit every year.

ILLUSTRATION 9 Ronda, Spain
PHOTO BY THE AUTHOR

Tourist meccas can be cities, such as Paris, London, Ronda (in Spain), or San Francisco; they can be areas of a country, such as Andalucía, in Spain; Normandy, in France; Cappadocia in Turkey; the Greek Islands, or the Pacific Northwest in the United States. They can be villages or small cities that hold special interest, such as Lourdes, in France or Perugia, in Italy or Granada, in Spain. In all cases, they are places that have something that attracts tourists. It could be due to any number of factors, such as its spectacular scenery, its religious significance, its great museums, or because it is a center of gay and lesbian or heterosexual sex. We visited Ronda while in Spain and there were so many tourists there you could hardly walk down the streets. It is a classic example of overtourism and is, like many other tourism meccas, too popular for its own good.

The game for many travelers is to find undiscovered gems and visit them before mass tourism begins and these places become over-developed over-touristed meccas full of tourist traps. The advent of mass tourism to a city or a place is a two-edged sword. On the one hand, when tourists come to a place they spend money on hotels, bars, restaurants, transportation, and entertainment and generally, do a good deal of shopping. On the other hand, cities don't want to be overwhelmed by huge numbers of other tourists who often damage the infrastructure and cause problems of one sort or another.

Tourism leads to new jobs and more money for the community, as a whole, though much of this money has to be devoted to upgrading the infrastructure and taking care of the needs and wishes of tourists. Tourism often leads to the escalation of the cost of living for natives and, in some cases, to the displacement of locals who, for example, may not have the easy access to beaches they had before beachfront hotels started springing up. Tourism frequently leads to the development of a sex industry and various kinds of culture clashes between tourists and natives. It turns out that some places, maybe even some countries, that have become tourist destinations become too dependent upon tourism, which then becomes the tail that wags the dog. The place, the area, and often the entire country become extremely dependent upon tourism, and attracting and satisfying tourists start becoming a dominant aspect of economic planning and social life.

3.14 *Uses and Gratifications of Tourism*

This analysis of the gratifications from tourism draws on a field of communications and social science research known as the "Uses and Gratifications" theory. Many years ago, social scientists started asking people why they listened to certain radio programs, watched certain television programs, went to certain films, or read certain kinds of novels. They weren't interested in the effects of the media but on how people used the media and the gratifications the media

ILLUSTRATION 10 Mezquita, Cordoba
PHOTO BY THE AUTHOR

offered them. From their answers, they compiled a list of gratifications that the media provided the general public, which led to a focus for many years in media and communication studies on "Uses and Gratifications Research." I will adopt some of their findings and apply them to the gratifications that visiting certain iconic tourist sites provide. As suggested earlier, one reason many people travel is that they have generally purchased so many things that they don't need very much else in the way of material goods and so they choose to travel to have interesting experiences and because they believe that travel, especially foreign travel, will broaden their perspectives on life and enrich their lives. That is a common theme in many books and articles on tourism, in general, and places of interest for tourists, in particular. Below, I offer a list of some of the gratifications that tourism provides. I start with our desire to see beautiful places.

TO SEE BEAUTIFUL PLACES

One of the pleasures of travel involves visiting beautiful cities, such as Paris or Venice, and natural locations such as the desert in Morocco, spectacular islands, and pristine beaches that we find spiritually uplifting and life-enhancing. Venice is popular because it is in such a beautiful city in a remarkable setting.

TO SATISFY OUR CURIOSITY

Humans are naturally curious and tourists are people who can travel to foreign lands to satisfy their curiosity about what life in other countries is like. Even tourists who visit foreign countries for short periods can gain insights into what is distinctive about the cultures they are visiting as well as gain a sense of perspective on life's possibilities. In many cases, tourists will have read articles and books about the places they are visiting and so have some background knowledge about the culture, along with lists of best restaurants and must-see places.

TO BE ENTERTAINED AND AMUSED

Tourists wish to have experiences that they will find entertaining and pleasurable. This might involve going to a great restaurant in a foreign country, attending a spectacle such as a bullfight in Spain, or watching a performance of folk dancing. One of the most important ways that tourists "entertain" themselves is by shopping for souvenirs—objects that will remind them of their experiences in foreign countries and which, as is sometimes the case, will be a bargain. Some of these souvenirs are given as gifts to others as an expression of affection or kept by the tourists to help them remember their experiences.

TO GAIN A SENSE OF THE FELLOWSHIP OF MEN AND WOMEN

Tourists usually derive a great deal of pleasure from interacting with people in other countries and gaining a sense of fellowship and community with them. Some tourism theorists suggest that some of these interactions, chance meetings that lead to "peak" experiences, are what tourists remember most and like best about their visits.

TO PARTICIPATE IN HISTORY

One thing that tourists gain when they travel is participating in history, which explains why they desire to be at important sporting events or to visit places where great historical events took place or great historical figures lived and worked. The French philosopher, Descartes, believed that "traveling is like conversing with men of other centuries." People want to connect to places and events of importance and through tourism—visiting Freud's home or the Alhambra in Granada (if you can get in) or the Blue Mosque in Istanbul or the Bastille—they can do this. This kind of tourism may represent an attempt to deal with the diffuse and subliminal anxiety many people have about the fact that, as J. William Fulbright once explained, our lives are minor events in the ongoing universe, minor events except for us, our families, and friends, that is.

TO OBTAIN OUTLETS FOR SEXUAL DRIVES IN A GUILT-FREE MANNER

Sexual tourism and sometimes sexual exploitation is an important element in the travel industry, even though governments in countries where it exists try (or claim they try) to curtail it. Many travelers feel that having sex outside monogamous relationships with the same sex, the opposite sex, or (unfortunately)

ILLUSTRATION 11 The Blue Mosque (Sultan Ahmet Camii), Istanbul
PHOTO BY AUTHOR

children in foreign countries is not morally problematic. Certain countries, such as Thailand, are famous for their sex industries. On Wikipedia we read:

> **Sex tourism** is travel to a different locale for the sake of sexual activity, particularly with prostitutes. The World Tourism Organization, a specialized agency of the United Nations, defines sex tourism as "trips organized from within the tourism sector, or from outside this sector but using its structures and networks, with the primary purpose of effecting a commercial sexual relationship by the tourist with residents at the destination."

It adds the following material on child sex tourism:

> Some people travel to engage in sex with child prostitutes in a practice called child sex tourism. While it is criminal in most countries, this multibillion-dollar industry is believed to involve as many as 2 million children around the world.
>
> https://en.wikipedia.org/wiki/Sex_tourism#Child_sex_tourism

Even if sex is not directly involved, much tourism is related to sex: couples going on vacation to have sex with their partners, to find partners, or to gaze at people of other cultures whose behaviors and dress arouse them.

3.15 *Tourism and the Consumption Imperative*

Most people are not aware of the size and importance of the tourism industry (now a multi-trillion-dollar industry) or the amount of research conducted by scholars in tourism schools in colleges and universities all over the world. There are dozens of scholarly journals devoted to various aspects of tourism, such are marketing, hotel management, eco-tourism, theories of tourism, and so on. As tourism asserts itself as a major economic force in the world, universities are now recognizing tourism's importance and devoting more resources to studying tourism.

There is another factor related to tourism as consumption that is of interest. According to the French sociologist Jean Baudrillard, there is an element of compulsion in tourism. Baudrillard explains that consumption must not be seen as a pleasure or a means of enjoyment but as a *duty*. In a fascinating reversal of the Puritan ethic of hard work, now enjoyment and consumption are the required "business" of the typical person who is, so to speak, now being forced to have fun. As Baudrillard writes in *The Consumer Society: Myths & Structures* (Sage, 1998: 80):

> There is no question for the consumer, for the modern citizen, of evading this enforced happiness and enjoyment, which is the equivalent in the new ethics of the traditional imperative to labour and produce. Modern man spends less and less of his life in production within work and more and more of it in the *production* and continual innovation of his own

ILLUSTRATION 12
Jean Baudrillard
DRAWING BY THE AUTHOR

> needs and well-being. He must constantly see to it that all his potentialities, all his consumer capacities are mobilized. If he forgets to do so, he will be gently and insistently reminded that he has no right not to be happy. It is not, then, true that he is passive. He is engaged in—has to engage in—continual activity.... Hence the revival of a *universal curiosity* ... You have to try *everything*, for consumerist man is haunted by the fear of "missing" something, some form of enjoyment. You never know whether a particular encounter, a particular experience (Christmas in the Canaries, eel in whiskey, the Prado, LSD, Japanese-style love-making) will not elicit some "sensation." It is no longer desire, or even "taste," or a specific inclination that are at stake, but a generalized curiosity, driven by a vague sense of unease—it is the "fun morality" or the imperative to enjoy oneself, to exploit to the full one's potential for thrills, pleasure or gratification.

Notice how travel and experiences of the differences found in various cultures play a part in Baudrillard's analysis. Christmas in the Canaries is a somewhat exotic location that might help the consumer get the pleasures that are due him, or, conversely, she can go to Madrid and visit the Prado, satisfying her need for culture and refinement in one of the great museums of the world. "Japanese-style love-making," whatever that may be, can be procured anywhere one finds Japanese men or women, but one must assume our consumer-tourist seeks as much authenticity as possible and goes to Japan to find the "real" thing—authentic Japanese-style lovemaking. Or is Japanese sex a better term? There is a driven aspect to all this, as if people, as consumers, feel they *must* suck the marrow out of life, and as tourists, must see everything worth seeing and go to every place that is worth visiting and eat everything worth eating. We are often animated by the fear of missing out (FOMO), which shapes our behavior in many different areas. The role of what we learn about foreign cultures and our imaginations envisioning them as "others" and, in extreme cases, "exotic," must be kept in mind.

Long before we go abroad to visit new lands, we become "virtual" tourists who imagine what the places we will visit are like, based on what we've seen in movies, travel advertisements, television documentaries, and read in magazines, newspapers, and guidebooks. But sometimes there are pleasant surprises.

When my wife and I went to Japan many years ago, we happened to stay, by chance, in an area of Tokyo when there was an important religious festival and parade, which really enhanced our visit to Japan. There was a festive quality to

ILLUSTRATION 13
Hooded Japanese women in parade
PHOTO BY THE AUTHOR

life, as people set up tents for drinking and partying and there were parades in which huge teams carried heavy religious statues around the streets. Many thousands of people had gathered around the temples and there was a sense of excitement that was contagious. It was as if we had been invited, suddenly, to a gigantic party.

3.16 *The Four Consumer Cultures and Tourism Preferences*

What follows next, is an analysis that suggests that the concept of a "consumption culture" is inadequate, for there are, it has been asserted by certain scholars, four—and not more than four—different consumption cultures or lifestyles in any society. Tourism, as we have just seen, is often linked with consumer culture as an example of the importance of consumption in modern society and of tourism's global significance. But saying that there is a consumer

ILLUSTRATION 14 Religious festival, Tokyo
PHOTO BY THE AUTHOR

culture is, it has been suggested, a bit of an oversimplification. There are, some scholars argue, not one but four different consumer cultures found in all societies, and these consumer cultures (or political cultures or lifestyles) are based on certain beliefs and values that members of each consumer culture hold in common.

This analysis draws upon the work of the political scientist Aaron Wildavsky and the social anthropologist, Mary Douglas, with whom he often collaborated. Wildavsky said that cultural theory tries to help answer two basic questions people face—one involves identity: *Who am I?* And the other involves action: *What should I do?* Wildavsky writes (in his article, "A Cultural Theory of Preference Formation" in Berger, 1989: 25):

> The question of identity may be answered by saying that individuals belong to a strong group, a collective that makes decisions binding on all members, or that their ties to others are weak in that their choices bind only themselves. The question of action is answered by responding that the individual is subject to many or few prescriptions, a free spirit or tightly constrained. The strength or weakness of group boundaries and the numerous or few, varied or similar prescriptions binding or freeing individuals are the components of their culture.

ILLUSTRATION 15
Mary Douglas
DRAWING BY THE AUTHOR

So there are groups with strong group boundaries and others with weak group boundaries. And there are groups with many rules and prescriptions and other groups with few rules and prescriptions. We end up with four groups—what Wildavsky calls "political cultures" or what Douglas calls "lifestyles":

CULTURE	LIFESTYLE
Strong boundaries and few rules	Egalitarians
Strong boundaries and many rules	(Hierarchical) Elitists
Weak boundaries and few rules	Individualists
Weak boundaries and many rules	Fatalists

There are, Wildavsky suggests, four political cultures—and, in the economic realm, as Douglas argues, four consumer cultures or lifestyles that arise from the answers to these two questions—what Wildavsky calls hierarchical, individualist, egalitarian, and fatalist cultures. Some grid-group theorists use different terms for each of the four lifestyles, I should add.

We can see how these cultures are formed out of the strength and weakness of group boundaries and the numbers and kinds of rules and prescriptions. As Wildavsky explains in Berger (Ed.) *Political Culture and Public Opinion* (1989: 25):

> Strong groups with numerous prescriptions that vary with social roles combine to form hierarchical collectivism. Strong groups whose members follow few prescriptions form an egalitarian culture, a shared life of voluntary consent without coercion or inequality. Competitive individualism joins few prescriptions with weak group boundaries, thereby

> encouraging ever new combinations. When groups are weak and prescriptions strong—so that decisions are made for them by people on the outside—the controlled culture is fatalistic.

Over the years, Wildavsky changed the names of some of his cultures, but the four I have listed give a good description of his ideas. He sometimes used Hierarchical Elitists or just plain Elitists instead of hierarchical collectivism, because he thought that description was a bit confusing. I should point out here, that members of these four political/consumer cultures generally don't recognize that they are members of one of these lifestyles and aren't aware of their existence. People have certain values and belief systems that Wildavsky and Douglas have identified as placing them in one of the four cultures.

What is important to recognize is that our membership in one of the four lifestyles shapes many aspects of our lives, including our choices for travel destinations and the way we travel. Mary Douglas suggests that it is their membership in one of the four consumer cultures or lifestyles—each of which is antagonistic or in conflict with the three others—that explains people's consumer choices. People may not be able to articulate their beliefs and values that put them in one of the four consumer cultures, but they recognize that their values and beliefs aren't those of members of the other consumer cultures. Consumption, then, is based on cultural alignments and hostilities and is not based on individual wants or desires. And tourist preferences, of course, are a kind of consumption.

Douglas explains how all of this works in the realm of consumption in her essay, "In Defence of Shopping" (in Pasi Falk and Colin Campbell, editors, *The Shopping Experience*, 1997: 19):

> None of these four lifestyles (individualist, hierarchical, enclavist, isolated) is new to students of consumer behavior. What may be new and unacceptable is the point that these are the only four distinctive lifestyles to be taken into account, and the other point, that each is set up in competition with the others. Mutual hostility is the force that accounts for their stability. These four distinct lifestyles persist because they rest on incompatible organizational principles. Each culture is a way of organizing; each is predatory on the others for time and space and resources. It is hard for them to co-exist peacefully, and yet they must, for the survival of each is the guarantee of the survival of the others. Hostility keeps them going.

She offers this theory to counter the theories of consumption that come from a framework based on individualist psychology and argues that "cultural alignment is the strongest predictor of preferences in a wide variety of fields." (1997:

23). By "cultural alignment," Douglas means our membership in one of the four lifestyles. This suggests that there is an inherent logic behind the shopping that people do and as Douglas argues, it is shoppers, consumers, whatever you wish to call them, who ultimately dictate what will be sold. We can see this is we examine the households of people from different consumer cultures. People in the different consumer cultures may have similar incomes (except for the fatalists/isolates, who are at the bottom of the income ladder, generally speaking) but the way they organize their households and their patterns of consumption, especially when it comes to travel, are shaped by their membership in one of the four consumer cultures.

Douglas concludes that the notion that shopping is the expression of individual wants is incorrect, for this notion doesn't take into account cultural bias. As she points out in the conclusion to her essay (1997: 30):

> The idea of consumer sovereignty in economic theory will be honoured in market research because it will be abundantly clear that the shopper sets the trends, and that new technology and new prices are adjuncts to achieving the shopper's goal. The shopper is not expecting to develop a personal identity by choice of commodities; that would be too difficult. *Shopping is agonistic, a struggle to define not what one is but what one is not.* [My italics] When we include not one cultural bias, but four, and when we allow that each is bringing critiques against the others, and when we see that the shopper is adopting postures of cultural defiance, then it all makes sense.

If Douglas is correct and there are four distinct and mutually antagonistic lifestyles—what I've called consumer cultures—then the various kinds and modes of travel would have to appeal to members of one consumer culture and not the other three. It wouldn't be socio-economic class and discretionary income that shapes travel decisions, but lifestyles or membership in one of the four consumer cultures, and, also, the desire to avoid members of the other lifestyles and consumer cultures. This means that members of each of the four antagonistic but mutually dependent consumer cultures or lifestyles look for different things when they travel and become different kinds of tourists. Let me hypothesize how people in each of the different consumer cultures travel, assuming Douglas is correct about antagonistic lifestyles.

ELITISTS search for hierarchy and obtain it when they travel by choosing luxury cruises or other luxury forms of travel, where they will find, they can assume—due to the cost of the travel—people like them. Luxury tours and everything associated with this kind of travel helps them maintain social distance from members of the other consumer cultures.

INDIVIDUALISTS, not wanting anyone to tell them what to do, will travel independently, booking their hotels and other aspects of their trips through their travel agents or on their own, using the Internet. They will avoid packaged tours but may take certain kinds of cruises that appeal to them.
EGALITARIANS will take tours that focus on cultural matters, nature preservation, and eco-tourism, wanting to do all they can to preserve nature and to show their links with people everywhere; they will avoid the kinds of tourism liked by Elitists and Individualists.
FATALISTS (who Douglas calls "isolates") and who are generally at the bottom of the financial totem pole and have little resources, will take trips by automobiles, buses, or short excursions. Those Fatalists who are relatively well off might take inexpensive cruises, of which there are many.

One reason why certain travel experiences may not be pleasurable for some people is that, by chance, they made the wrong kind of travel arrangements and ended up with people from a different and antagonistic lifestyle or consumer culture.

Wildavsky and Douglas force us to consider that the various analyses of travel and tourism as a kind of consumption made by scholars unaware of the importance of the different consumer cultures, may be overly simplistic and inadequate. If we want to know why people choose to travel the way they do, according to Wildavsky and Douglas, we shouldn't probe their minds and psyches but find out to which lifestyle/consumer culture they belong. That is because, as Douglas explains, it is "cultural alignment" that explains preferences in politics and tourism and all other forms of consumption, including shopping—which is one of the main things people do as tourists. Shopping, along with a search for authenticity, motivates many tourists, though as I've explained above, there are many different kinds of tourists and motivations that lead people to become one or another kind of tourist.

3.17 *The Search for Authenticity*

One of the major debates in the field of tourism studies involves the matter of the tourist's search for authenticity. As Dean MacCannell writes in *The Tourist* (1976: 145):

> I have found the solidarity of modernity to be grounded in objective relations, in history, and social facts. But unique to the modern world is its capacity to transform material relations into symbolic expressions and back again, while continuing to differentiate or multiply structures. The

> expansion of alternative realities makes the *dialectics of authenticity* the key to the development of the modern world. The question of authenticity transcends and subsumes the old divisions of man vs. Society, normal vs. Deviant, worker vs. Owner.

So authenticity emerges, for MacCannell, as a seemingly all-powerful desire in people and, particularly in tourists, to see life "as it is really lived." (Other tourism scholars, I should add, disagree with MacCannell about the importance of authenticity.) Is the search for authenticity the basic reason for tourism? Daniel Boorstin, in his book, *The Image*, argues that modern tourists tend to be shallow individuals who are essentially in search of entertainment and are satisfied by the fake "pseudo-event" experiences sold to them by the tourist industry. As Boorstin writes (1961: 114):

> The traveler ... was working at something: the tourist is a pleasure-seeker. The traveler was active: he went strenuously in search of people, of adventure, of experience. The tourist is passive: he expects interesting things to happen to him. He goes sightseeing.

ILLUSTRATION 16 Jewish graveyard, Poland
PHOTO BY THE AUTHOR

Boorstin offers many distinctions between travelers and tourists, but there is some question about how valid they are. We can see Boorstin's ideas about the differences between travelers and tourists in the following chart, taken from his writings on the subject quoted above.

TRAVELERS	TOURISTS
Working at something	Pleasure seeking
Active	Passive
Searches for adventure	Expects things to happen to him
Wants authenticity	Wants pseudo-events

Boorstin is oversimplifying things to a considerable degree and is an elitist who looks down on most people. In *The Tourist*, MacCannell suggests that Boorstin is expressing a typical feeling that tourists have—of dislike for *other* tourists. Tourists are caught up in a fundamentally ironic situation—they want to go places that are interesting and exotic but where there aren't other tourists or so many tourists that visiting a place is unpleasant. Or, perhaps, as my discussion of consumer cultures suggests, where they can avoid tourists who are not like them. This places a premium on discovering new "in" or "hot" places to go before mass tourism arrives.

MacCannell argues that modern tourists are seeking authenticity in other lands as a means, among other things, of gaining insights into their own lives. He agrees, however, with Boorstin's notion that the tourist industry provides staged or inauthentic experiences, and fake history that gullible tourists, searching for authenticity, accept as real. Both Boorstin and MacCannell seem to believe that tourists are passive, gullible, and easily fooled by the ersatz, fake, pseudo-experiences that the travel industry provides them. There is no proof that this is always—or even often—the case, and while there are many "tourist traps" and while some tourists are gullible, to generalize the way Boorstin and MacCannell do is difficult to defend. Many tourists, I would suggest, can recognize the difference between real and staged experiences, and others—if the postmodern theorists are right—don't care very much about whether their experiences are "real" or "fake." They want to have fun and see both kinds of experiences as part of what they are looking for, which might be described as entertainment and amusement.

The fact of the matter is that tourists are attracted to places that they find beautiful, that have cultural sites of interest, and that help them escape from the patterns they follow in their regular or everyday lives.

Generally speaking, if a place is full of tourists it is because there are attractions worth seeking out. I can recall a short bus trip my wife and I took during a recent visit to Puerto Vallarta. It was to a town that shall be nameless—a

ILLUSTRATION 17 Bali dance show
PHOTO BY THE AUTHOR

forty-five-minute bus ride from Puerto Vallarta. We were the only tourists on the bus, and for good reason. And that is because there was nothing of interest in that town at all. It was a small nondescript town with nothing that would attract anyone unless you wished to see an "authentic" small and uninteresting Mexican town at the end of a road to nowhere.

So authenticity isn't enough—you have to have something authentic that is worth seeing. Thus, there are two extremes facing tourists: the inauthentic "staged" fake native culture that Boorstin writes about and the authentic but uninteresting "real" culture that I discovered at the end of our long bus ride to the Mexican town. But there are a lot of places between these two extremes where tourists find pleasure and satisfaction. It's very much like eating in a crowded restaurant. There usually are good reasons why some restaurants are crowded and others are empty.

3.18 *In Quest of the Exotic*

One of the dominant goals for many tourists is finding what might be called an exotic destination. The term "exotic" comes from the Greek word *exotikos*, which means "outside." For tourists, the exotic means some combination of remote, unusual, different, foreign, charming, fascinating, enchanting, and distant in time. To the extent that tourists are searching for something different from their everyday lives, the exotic represents the most distant polarity.

In the popular mind, I would suggest that certain so-called primitive cultures and various South American, African, and Asian cultures are seen as exotic by Americans. The French may be different from us in many respects, but they are not exotic, whereas certain people in some parts of Africa, India, Southeast Asia, and various other places like them, where matters such as dress, language, cuisine, and architecture differ greatly from American society, are seen as exotic. We can see the difference between our everyday lives and the exotic in the following comparison:

EVERYDAY LIFE	THE EXOTIC
Near	Far
The present	The past
Familiar	Strange
The Supermarket	The Souk
Cathedrals	Hindu temples, Mosques
Euro-American cuisine	Ethnic cuisines
Suits, dresses	Turbans, robes, caftans

There are, of course, different forms of the exotic, but generally they involve some combination of that which is strange to us, distant in place and time from us, and traditional rather than modern, as these phenomena apply to such things as landscape, architecture, dress, food, languages, and cuisines. In Vietnam, the little girls who clustered around us in Sapa, in their traditional Black Hmong and Red Zhao costumes, were exotic. They spoke four or five languages and they dressed in native costumes that hadn't changed in hundreds of years. Bali is an exotic destination: beautiful and with a fascinating culture. It seems likely that tourism has led to the resurrection of many ancient dances, so tourism is playing a positive role, in some respects, in Bali. Tourism also has negative effects there as well, as might be expected, and Bali is suffering from overtourism. Sometimes, you find a mixture of the exotic and the ordinary in the same country. In Morocco, for example, a city such as Casablanca is European, and therefore similar to our everyday experiences of contemporary cities, while some places in Morocco are quite exotic, like the souk in Fez or the Place Djemma el Fna in Marrakech.

3.19 *Travel as a Rite of Passage for Young Men and Women*

Several hundred years ago, young men from wealthy families in Europe often took what was known as "The Grand Tour" in Europe, and in some cases, elsewhere. It was held that travel would give young men a certain amount of polish and sophistication, as a result of their spending time in some of the great cities and visiting important museums and other cultural venues. The Grand Tour was, obviously, part of an elitist "Rite of Passage." Women were not traditionally sent on the Grand Tour because, in those days, women were supposed to be "sheltered" and because that tour was not felt to be of importance to them in their future roles as mothers and homemakers.

This Grand Tour no longer is reserved for the elites, and now legions of young men and women from countries all over the world travel everywhere. For example, American high school students frequently travel to Europe during summers and many college students now take their junior year abroad. It is a modernized version of the Grand Tour that has been democratized and massified, but the purposes are the same—to broaden and educate young men and women by exposing them to other cultures. They visit great museums, they experience other cultures, eat different foods, and have adventures. Also, it is felt that traveling abroad and, even better, living abroad, helps young men and women gain an appreciation of American culture. In fact, among many middle-class students in America, going abroad is not seen as something remarkable but rather is taken for granted. Now it is *not* traveling abroad that is considered surprising and those who do not travel tend to be seen as, in certain ways, deprived.

Travel abroad now seems to be seen as part of the transition from youth to adulthood; during travel, one is in what anthropologist Victor Turner calls a "liminal" state. He defines liminality in his book, *Dramas, Fields, and Metaphors: Symbolic Action in Human Society* (Cornell, 1974: 232) as "ambiguous, neither here nor there, betwixt and between all fixed points of classifications." The foreign travels of high school and college students are then, ritual activities tied to their liminality, which marks the period in which their childhood is ending and young adulthood is dawning. After their travels, these students will generally be re-integrated into their communities and societies but will have a different status. Rites of passages are generally "sacred" in nature and are connected with important changes in people's lives and usually help, in this case (or so it is hoped) with their transition into adulthood.

3.20 *An Imaginary Image that Explains Things*

To visualize how complicated the travel industry is, imagine, in your mind's eye, a person—a tourist—standing in the middle of a large room. Circling around that tourist are people who do certain things or represent certain industries related to travel. I have mentioned some of these matters earlier. I refer to cab drivers, bus drivers, train conductors, airline pilots, cruise ship captains, clothing manufacturers, travel agents, tour company representatives, hotel clerks, local guides, guidebook publishers, owners of restaurants, museum docents, symphony orchestra musicians, dancers and singers from night clubs, players from baseball teams and other sports teams, clerks from gambling casinos ... the list goes on and on. So when Jane Doe calls up her travel agent and says she and her husband want to book a two-week tour of Spain, that decision is connected to many kinds of people and institutions, everywhere. In a certain sense, travel is simple—but that's because so many institutions, in various places, all over the world, are available to take care of the needs and desires of travelers and tourists. The trip is the tip of the iceberg. Underneath that trip are all the people and companies and institutions that make the trip possible.

4 The Mind Set of the Traveler

> In a vast portion of human history, men have been the travelers; and travel literature is—with a few significant, and often modern exceptions—a male literature reflecting a masculine point of view. This comes as no surprise to those who have digested the insistence of modern feminists that men have generated and monopolized representational realities in which the voice of women is silent, undercut, or assumed. In the history of patriarchal civilizations—and, as yet, there are no other kinds—humanity has worn the mask of masculinity, and travel has been a performance of this persona. The masculinization of motion and the feminization of sessility [being attached] are clearly products of cultural patterning, many examples of which occur through the travel literature. In patriarchal cultures, the mobility of men—especially of young, unattached men—is overdetermined and powerfully charged by the reigning images of masculinity, whether of the wandering knight or of the wandering holy man, the shaman, or the actor.
>
> ERIC J. LEED, The *Mind of the Traveler: From Gilgamesh to Global Tourism*. New York: Basic Books, 1991

> Students of tourism have for long been oblivious to the significance of this rising trend for the sociological analysis of tourism. We know little about the motivations and desires of the growing numbers of non-Western tourists; but it is reasonable to assume, that they did not share the Westerners' "art of travel" in the past; nor are they necessarily impelled by a quest for "authenticity" at present. The question, what are the culturally justified motives for travel of non-Western travelers, and how these help to fashion their style of travel, and their choices of destinations and activities, is one of the major issues, presently facing the sociological study of tourism.
>
> ERIK COHEN, "The Changing Faces of Contemporary Tourism". *Society*. Vol. 45, No. 4, July/August 2008

There are other aspects of tourism that are worth considering, such as the fact that, to varying degrees, travelers are risk-takers and strangers in strange lands, and much travel is fueled by nostalgia for the past.

4.1 *Travelers as Risk-Takers*

Travel always involves elements of "risk," since things often go wrong when you travel—your flight is delayed and you miss your connecting flight or the bus you are traveling on breaks down. The things that go wrong are generally inconveniences, not serious risks, but the fact remains that it takes a certain kind of personality to travel. You have to expect that, for one reason or another, something might go wrong and be willing to deal with the problems that arise. Many people are risk-averse and either do not travel or find ways of traveling, such as taking group tours that minimize risk. The German sociologist Georg Simmel was interested in what he called "adventures." He explained that an adventure involves dropping out, for a precise period, of the continuity of life, and abandoning the regularities and certainties that are part of our everyday lives. To me, that is a pretty good definition of being a tourist. Simmel added that adventures take on the qualities of dreams in our psychic life and they have much sharper beginnings and endings than other aspects of our experience, which helps explain why we remember them so vividly. It is this capacity to accept the role of risk and chance in our lives that connects the traveler and the adventurer. And because our travels and adventures are so exhilarating, we remember them quite vividly.

4.2 *Travelers as Strangers*

Simmel also pointed out that strangers—people who are new to someplace—tend to see things that don't attract the attention of natives or are almost invisible to natives. Strangers can be more objective than natives. That's because the reference points for strangers are new and the customs and traditions which are part of the lives of natives are not familiar to strangers and thus, of some interest. The natives are so used to things that they pay no attention to behaviors, rites, and customs that, to strangers (such as anthropologists and tourists) are often quite remarkable.

So travelers are people who put themselves in the situation of being—if only for short periods—strangers and who derive enjoyment from this status. Being a stranger means travelers have to be more active, have to find ways to make sense of "foreign" things, have to figure out how to take care of basic necessities, and this has certain psychological benefits in making the traveler feel more alive and resourceful. This is only possible where the place visited is not too far removed from what the traveler is used to; if travelers don't know the language spoken and know nothing about the culture at all, they need help—from guides and others who see to it that their needs are taken care of. Being a stranger has its difficulties, but it also has its pleasures and that pleasure—of investigating that which is strange and seeing things for oneself—is one of the strongest motivations for travel.

4.3 *Travel Fueled by Nostalgia*

One element of travel, which often involves going to sites of great historical interest, can be explained as a kind of longing for the imagined wonders of the past and wondering what travel was like during this imaginary past. Thus, the great French anthropologist, Claude Lévi-Strauss, laments that he didn't have a chance to experience "real" travel. As he writes (*Tristes Tropiques: an anthropological study of primitive societies in Brazil.* New York: Atheneum. 1970: 44, 45).

> I should have liked to live in the age of *real* travel, when the spectacle on offer had not yet been blemished, contaminated, and confounded; then I could have seen Lahore not as I saw it, but as it appeared to Bernier, Tavernier, Manucci.... There's no end, of course, to such conjectures. When was the right moment to see India? At what period would the study of the Brazilian savage have yielded the purest satisfaction and the savage himself have been at his peak?.... The alternative is inescapable: either I am a traveler in ancient times, and faced with a prodigious spectacle which would be almost entirely unintelligible to me and might, indeed, provoke me to mockery or disgust; or I am a traveller of our own day, hastening in search of a vanished reality.

There is this nostalgia for the days when there was such a thing as "real" travel and travel hadn't been corrupted by the mass tourism that haunts Lévi-Strauss's book. *Tristes Tropiques* is now considered a classic travel book, though he announces in his book that he loathes travel. Bill Buford, who edited a collection of travel articles published in *Granta* magazine, commented about the effect of modernization on travel, lamenting that the Second World War marked the end of "travel" as we used to know it and the beginning of mass tourism. Buford writes (in *Granta: Travel Writing*, 1984):

> World War Two marked the end of the great age of travel. And it is easy to see why. Not only did the war make travel difficult, but the world that emerged after it—a monoculture of mass consumerism, package-holidays, and the extraordinary imperialism of the American hamburger—left little room for the sensibility characteristic of so many

ILLUSTRATION 18 Taj Mahal, India
PHOTO BY THE AUTHOR

> of the writers of the twenties and thirties. The books of Robert Byron, say, or Waugh, Auden, and Priestly are virtually Edwardian in temperament: they hark back to an earlier time. They wrote, as Paul Fussell has pointed out in *Abroad*, to celebrate a "Golden Age": the corners of the world uncontaminated by the manifestations of modernity. "One travelled," Fussell says, "to discover the past."
>
> https://granta.com/editorial-travelwriting

This "past," of course, was one enjoyed by the elites. Soon the world was to be "contaminated" by the gastronomic imperialism of the American hamburger and the development of mass tourism, forcing the world to put up with the horrors of hordes of "ugly" middle-class Americans.

4.4 *Travel and Ambivalence towards Modernity*

Connected to this nostalgia for the possibility of "real" travel in the past is a powerful nostalgia for what I would call the "imaginary" past itself when life seemed to have been more interesting and exciting.

Dean MacCannell touches on this matter when he writes in *The Tourist*: (1976: 2, 3):

> Modernity first appears to everyone as it did to Lévi-Strauss, as disorganized fragments, alienating, wasteful, violent, superficial, unplanned, unstable, and inauthentic. On second examination, however, this appearance seems almost a mask, for beneath the disorderly exterior, modern society hides a firm resolve to establish itself on a worldwide base. Modern values are transcending the old divisions between the Communist East and the Capitalist West and between the "developed" and "third" worlds. The progress of modernity ("modernization") depends on its very sense of instability and inauthenticity. For moderns, reality and authenticity are thought to be elsewhere: in other historical periods and other cultures, in purer, simpler lifestyles. In other words, the concerns of moderns for "naturalness," their nostalgia, and their search for authenticity are not merely casual and somewhat decadent, though harmless, attachments to the souvenirs of destroyed cultures and dead epochs. They are also components of the conquering spirit of modernity—the grounds of its unifying consciousness.

This unifying consciousness is located, for MacCannell, in the middle classes, whose members, as he describes the situation, scavenge "the earth for new experiences to be woven into a collective, touristic version of other peoples and other places." (1976: 13) His language suggests he does not think much of the behavior of the middle classes. In the past, only the wealthy or aristocrats traveled because they were the only ones who could afford to travel. Mass tourism is now essentially a middle-class phenomenon and, for that reason, is attacked by elitists and others who, it would seem, identify either with aristocratic philosophical beliefs or Marxist and left-wing ideological beliefs. These two groups are united in despising the middle classes, and their nostalgia is tied either to a dream of the past, of an aristocratic society when there were no middle-class people, or of the future, and a utopian classless, all middle-class world.

4.5 *Travel as a Means of Personal Transformation*

Travel offers people the ability to transform themselves in varying ways. It is generally held that travel "broadens" people by allowing them to see how others live and to experience great works of art, which it is commonly believed "elevate" people both spiritually and emotionally. So, the notion that travel can "change" people is commonly held.

But there is a good reason to believe that travel also enables people to transform themselves in more profound ways. In our everyday lives, we become locked, so to speak, into a series of relationships that offer us our identity. We gain our sense of who we are by the way others react to us, and sometimes, as a result of chance events and unfortunate circumstances, we obtain an identity that is not one we like. It is significant others, at our jobs, and in our social relationships who confer identity upon us. One thing that travel does is enable us to escape from the identities we have been given and try on new identities. Paul Fussell explains this in a book he edited, *The Norton Book of Travel.* New York: W.W. Norton (1987: 13):

> Why is travel so exciting? Partly because it triggers the thrill of escape, from the constriction of the daily, the job, the boss, the parents. "A great part of the pleasure of travel," says Freud, "lies in the fulfillment of … early wishes to escape the family and especially the father." The escape is also from the traveler's domestic identity, and among strangers, a new sense of selfhood can be tried on, like a costume. The anthropologist Claude Lévi-Strauss notes that a traveler takes a journey not just in space and time (most travel being to places more ancient that the traveler's home) but "in the social hierarchy as well"; and he has noticed repeatedly that arriving in a new place, he has suddenly become rich (travelers to Mexico, China, or India will know the feeling). The traveler's escape, at least since the Industrial Age, has also been from the ugliness and racket of Western cities, and from factories, parking lots, boring turnpikes, and roadside squalor.

So travel provides escape from both personal identity and if one goes to certain countries, class identity. When you travel to some countries—such as Vietnam and a host of other lands in Asia and Southeast Asia, you become, for a short period, relatively "rich" and live the way the rich do. What this escape from the "social" hierarchy or status one is caught in during one's everyday life is to rupture the old identity and make it possible for travelers to assume a different identity when they return to their everyday routines. Travelers can try on their new identities "like a costume" and if they like them, adopt them, with necessary modifications, when they return to their regular lives. This would suggest that travel, especially travel in foreign countries can have rather profound effects upon people, which would explain why so many people find themselves drawn to foreign travel. They may not fully recognize the benefits they derive from this kind of travel but they do sense that their travel has been a valuable experience.

Psychoanalytic theorists talk about the process of separation and individuation in individuals. According to this theory, people must find ways to separate themselves from the dominating influence of their mothers and fathers and find a way to achieve an identity of their own. This is similar to what anthropologists talk about when they write about separation, liminality, and achieving an identity. The two theories are similar in that there is some kind of a break with one's family and with one's childhood, a liminal period in which one is "in-between" childhood and adulthood, and a resolution in which a person emerges changed considerably. I am suggesting that travel may be something like these two processes. One escapes, if only for a short period, from the life to which one is used. This temporary status that the tourist experiences can be described as being a liminal period. During the trip, one is "educated" about the different possibilities that exist for living one's life and one's horizons are broadened.

At the end of the trip, it is quite likely that travelers have undergone experiences that change, to varying degrees, their attitudes, beliefs, and values. Some people have told me that after some tours they have taken their lives were changed in very positive ways. There is always the hope that a trip will be transformative and that after we travel, we see the world a bit differently than the way we saw it before we left and we will see ourselves differently, as well.

5 Notes on Ocean Cruising

> Edward Stephan, who died November 8 at the age of 87, was among the first entrepreneurs to see the vast potential of Miami as a base for cruises in the Bahamas and Caribbean. Founding what is now called Royal Caribbean Cruises Ltd. In 1969, he joined two other outsiders—Knut Kloster of Norway and Ted Arison, an Israeli—in persuading middle-class Americans that cruises weren't just for the idle rich, or the "overfed and nearly dead," s one of Mr. Stephan's aides put it. Serving a mass market that didn't exist in the 1960s, Royal Caribbean is now the world's second-largest cruise operator, after Carnival Corp, founded by Mr. Arison. Norwegian Cruise Line Holdings Ltd., founded by Mr. Kloster, is third.
>
> JAMES A. HAGERTY, "Entrepreneur Sold Cruises to Middle-Class Americans". Obituaries. *The Wall Street Journal.* Nov. 17–18, 2019

The growth rate for the cruise industry tops just about anything in the travel world. Last year, 28.5 million people took a cruise, nearly 7% more than in 2017, according to the Cruise Line International Association. The

> growth rate in North America was 9%. A total of 19 new ocean trips are expected to sail in 2020. CLIA says global passenger count should reach 32 million.
>
> SCOTT MCCARTNEY, "Making 180,000 tons Feel Cozy at Sea." *The Wall Street Journal*. Dec. 19, 2019, page A11

We move now to one form of travel, ocean cruising, which has become one of the most popular forms of travel. I would describe this discussion as ethnographic and semiotic in nature. At one time, cruises were described as essentially "for the newlywed and the nearly dead," but that doesn't apply to the industry now. Of course, cruise lines differ. The average age of passengers on Carnival lines is much younger than passengers on Holland America lines. The average age of the passengers on a Holland America cruise we took must have been over seventy. It was a long cruise and generally speaking, the longer the cruise, the older the passengers. Here is some information on this matter:

> According to the Cruise Lines International Association (CLIA) Global Passenger Report, the average age of cruise passengers was 46.7 years old—holding steady from the previous two years. However, 40 to 49-year-olds make up only 15% of all cruise passengers for that year. This is probably due to a handful of families with small children lowering the average

ILLUSTRATION 19 The *Norwegian Epic*
PHOTO BY THE AUTHOR

> age. The median age was between 60 and 69-year-olds, with a full 19% of cruisers falling under this demographic.
>
> https://www.cruise1st.co.uk/blog/cruise-holidays/how-old-is-the-average-cruise-passenger

5.1 *Aren't Cruises Boring?*

At a party recently, I got into a conversation with a woman who had worked as a cardiology researcher at the University of California Medical School before she retired. She was married to a man she had met in England, where she did her postgraduate work. He also worked at the medical school. I believe she was a physician and her husband had a Ph.D. or vice versa. She mentioned that they often traveled to Western Europe for their vacations. I told her that my wife and I were going to be taking a cruise to Mexico. "Aren't cruises boring?" she asked, adding that she had never taken a cruise and that she hadn't ever been to Mexico. In the popular mind, cruises are full of people who spend most of their time eating and who don't do much in between meals. "Isn't Mexico dangerous?" she said. Many people fear traveling in Mexico because they read about the gang killings there. I'd say that going to Mexico is as dangerous for tourists as walking around downtown San Francisco.

Statistics show that only around fifteen percent of Americans have taken cruises, and many members of that group take many cruises throughout their careers as travelers. Obviously, these people, who may take dozens of cruises, don't find it boring. So why do people ask "aren't cruises boring?" The question we must consider is what we mean by the term "boring?" Google lists some synonyms for boring: tedious, dull, monotonous, repetitive, unvaried, and uneventful.

We tend to use the term "boring" for things we don't like or think we don't like or think we wouldn't like, such as taking a cruise. There are many things to do every day and every night on cruise ships. I like to swim in the pools and soak in the hot tubs. And my wife and I go to all the shows in the cruise ship theaters. Many mornings my wife and I take a walk on the teak "walking deck" along with many other passengers. (Many of the walking decks are no longer teak, perhaps because it is so expensive to install and maintain them.) There are exercise classes every morning, there are lectures many days by people brought aboard to talk on various topics, there may be string quartets or string trios playing classical music, and at night there are entertainments—of varying quality, including a couple of big shows with singers and dancers.

There are usually bridge games every day full of contract bridge enthusiasts. And there is a casino for people who like to gamble. And, of course, if you wish to read a book you can do so in many places on the ship. And there is the ship to explore. Just wandering around the ship and learning where the dining rooms are, where the library is, or the location of the gym, can occupy one for a while. Cruise ships offer an abundance of things to do for passengers—so many that it is a bit overwhelming. And yet, of course, life has its routines—breakfasts, lunches, dinners, high-tea in some ships (usually around 3:30 PM) and food, always available in the cafeterias and assorted pizza shacks and hot-dog and hamburger restaurants on the ships.

My point is that life, by its very nature, is based on routines that we repeat over and over again, and it is only when life isn't routine or boring that there is a reason to be anxious. Cruise ships provide us with so many things to do (if we want to do them) that we become paralyzed by all the activities and often find it difficult to decide what to do. Many people enjoy just lying around the pools, catching some sun, and reading, chatting, drinking—whatever. I suggested to this woman, with whom I was talking, that she consider taking a cruise. "You can take a cruise from San Francisco and go to Mexico," I said. "Two firsts for you and very easy to do. You live in Sausalito. You can grab a cab to the pier in San Francisco and be there in half an hour. You can also cruise to Alaska from San Francisco but it wouldn't be that different for you, though the scenery is fabulous." "Actually," she said, "that sounds like a good idea. We've seen the commercials for river cruises. They look interesting and are probably less crowded than ocean cruises."

"Not necessarily," I replied. "Two hundred people on a small boat can be much more crowded than three thousand people on a very big boat. And if you're worried about not liking the kind of people who take cruises, you can always take a cruise with another couple or two, and that way you don't have to worry about whether or not you'd like dining with your fellow passengers. And you can choose, during meal hours, to dine whenever you want—on Princess ships it is called 'Anytime Dining'—instead of being at the same table every evening with people you may not find interesting." And that is where we left things. I found it quite remarkable that someone could live in the San Francisco area for twenty-five years and never have visited Mexico. Cabo is only a three hours flight from San Francisco and Puerto Vallarta is only three and a half hours away.

Taking a cruise can be boring if you refuse to take advantage of all the activities (line dancing, ballroom dancing, cooking shows, games, etc.) but it doesn't have to be. Boredom, like beauty, is in the eye of the beholder. I should add, that cruise liners stop in various ports during the cruise, so there are places to explore off the ship. On a recent cruise to Mexico we spent three days at sea, and then visited Cabo San Lucas, La Paz, Loreto, and Puerto Vallarta. At each port, Princess offers a dozen or so excursions. In Cabo, there are excursions for all tastes. What follows is an example of some of the more interesting ones:

$69 Cabo Resort Break
$99 High Sierra Zip Line
$69 Beach Horseback Ride
$99 Desert and Camel Adventure
$199 Desert Rhino
$89 Whale Watching
$149 Swim with Dolphins

And there are many, many other excursions in Cabo and all the other ports. Princess lines (and all cruise lines) make a great deal of money on these excursions and we received many emails from Princess over the weeks before we sailed, telling us that time was running out and if we didn't book one of these excursions now, we might not be able to take it. However, we never take excursions. If we want to visit some area we've never been to before, we hire a cab and have the driver give us a tour. In many ports, locals offer tours that are much less expensive than ship excursions. If you don't take excursions and are careful about your drinking and avoid gambling, cruising can be quite inexpensive. On the other hand, if you take an excursion at every port at $69 per person, you'll spend around $280 for excursions, which means an inexpensive cruise starts getting more expensive. And most excursions cost much more than 69 dollars.

But even if you take excursions at every port and drink and gamble a bit, the Mexican cruise on the *Star Princess*, for example, is still inexpensive in comparison to ocean cruises on upscale lines like Crystal or Regent Seven Seas or river cruises in Europe. And you have to fly to Europe for river cruises which is tiring and expensive. Assuming a flight to Europe costs $1000 a person and the river cruise costs $300 a day per person or $6000 for a couple on a ten-day cruise, we can see that river cruising is relatively expensive and a ten-day cruise costs $8000 and up. You get free excursions and drinks and better food, but you end up paying a great deal more than ocean cruises.

5.2 *Social Class and Cruising*

Cruise ships do not, as a rule, have different classes of passengers the way ocean liners once did, but different cruise lines cater to different classes of people. The Princess Line and many other lines such as Norwegian and Royal Caribbean cater to the middle-class cruise takers (the average salary of cruise takers is over $100,000 a year) while some lines, such as Regent Seven Seas, are much more expensive and cater to wealthy upper-class cruise takers or less wealthy people who are "splurging.". Now many cruise ships, such as the *Norwegian Epic*, have certain parts of the ship set aside for people who pay more than the typical passenger for special cabins and benefits. So social class and class differentiation are returning to many mass tourism cruise liners.

A 10-day cruise on Regent Seven Seas can cost a thousand dollars a day for the least expensive cabin. That's more than my wife and I paid for a 10-day cruise we took from San Francisco to Mexico and back (not including tips and taxes) on the *Grand Princess*. Other cruise lines vary in prices from what Princess charges to what Regent Seven Seas charges. Below I have copied material from a Regent Seven Seas brochure about a ten-day cruise they offer from October 23, 2021, to November 2, 2021:

> ROUND TRIP TOKYO
> **Seven Seas Explorer**
>
> (Included Business Class Air)
>
> $54,999 per person Regent Suite
> $19,399 per person Distinctive Suites
> $15,799 per person Penthouse Suites
> $13,999 per person Concierge Suites
> $12,799 per person Veranda Suites

That cruise would cost a couple around $13,000 for the least expensive cabins on the ship. Seven Seas is an ultra-luxury line and you get free drinks, free tours, and many other benefits. But for the same amount of money, about $1200 a day, my wife and I can take a 10-day cruise to Mexico. My wife and I are booked on a 10-day round trip cruise from San Francisco to Mexico on the Ruby Princess on November 27, 2021, which will cost us $750 per person for an obstructed ocean view cabin. When you add the port taxes it comes to about $1000 a person to which one must add $150 per person for tips.

What is remarkable about the Princess cruises and other premier cruise lines is that the quality of the food is pretty good in the dining rooms and the service is generally superb. A friend of mine, who took many cruises, explained to me that "dining on a cruise ship is like going to a decent French restaurant. And it's free."

So, once you pay for the cruise and the tips, which are deducted automatically, there are no other expenses, except ones you incur if you gamble or drink or take excursions. People generally spend around $500 per person for extras on a seven-day cruise according to a commentator on the Internet, who has a series of travel tips on cruising. There was a notice on the Internet on July 28, 2019, from the *New York Times* by its Frugal Traveler, Elaine Glusac, about "A Three-Day Cruise to the Bahamas for $210? There Must be a Catch." We read, "Budget cruises abound these days. For the price of a one-night stay at a moderately priced hotel, we decided to check one out." The article describes her three-day cruise and suggests that cruises can be less expensive, and more relaxing, than road trips and stays in hotels.

https://www.nytimes.com/2019/07/26/travel/bahamas-budget-cruise.html

A cruise liner can be thought of as a resort on water, and now that one ship has a roller coaster on it, also an amusement park on the water. The cruise experience is similar to going to an all-inclusive resort—except that on cruise ships you usually have to pay for your drinks. When we are on cruises, we only dine in the dining rooms, unless they are closed for lunches. Some passengers prefer the cafeterias on cruise ships but we avoid them to the extent we can. The breakfasts in the dining rooms of all the cruises we've been on have been wonderful. I describe the breakfasts on cruise ships as "hyper-baconized," because the amount of bacon these ships serve for breakfast is astronomical. On a typical large cruise ship, there are many restaurants to dine in. In addition to the free cafeterias and dining rooms, there are often specialty restaurants that cost extra.

5.3 *Statistics on the Cruise Industry*

There has hardly been a day in the past few years in which I haven't received an email from some cruise line offering fantastic "deals" and several times a week we receive brochures from cruise lines. It must be a very profitable business because the cruise lines spend a great deal of money sending expensive catalogs and brochures to me or my wife, and advertising in newspapers, magazines, and on television.

These statistics gathered from Statistica and the CLIA: Cruise Lines International Association websites, offer us some interesting information about the economic impact of the cruise industry. The cruise lines will be building 47 new ships between 2015 and 2022, which suggests that cruising is popular and growing increasingly so. Here is some information on the cruise industry before the pandemic:

$35 billion	World-wide cruise revenues
$1700	Average revenue per person
$17 billion	North American revenue
14 million	Number of cruise passengers in the USA

185	Ships in the American cruise industry
$114,000	Average income of cruise takers (approximate)
69%	Cruise takers with a college education
84%	Cruise takers who are married
49 years	Average age of cruise takers

The CLIA made a list of benefits cruise takers thought were important. This list follows and contains no surprise:

70%	Visit several ports on one trip
60%	Relax, get away from it all
60%	Be pampered
59%	High-quality entertainment
55%	Fine dining

There were many other benefits in the list but the ones above are among the most important benefits of cruising. This information was found by surveys. These data cover ocean cruising and river cruising. Fine dining is an important benefit on a cruise and meals in the dining halls take a considerable amount of time—in part, because of the number of courses served. The more people at a table, the longer the meals last, since it takes some time to serve everyone. A good deal of time during meals is spent with conversations.

The conversations we have when we eat in the dining rooms have a ritualistic quality to them. After everyone introduces themselves, people talk about whether they are enjoying or not enjoying the cruise, about ports they like or dislike, about the food on the ship, about where their next cruise will be, and about how many cruises they have taken over the years. You might ask why it takes two hours (or more) to have dinner on a cruise. The menu below explains why this is the case. The menus on cruise ships are very elaborate and imitate the kind of menus you might find in a fancy French or Italian restaurant. It is amazing how the cooks can feed three thousand people with relatively high-quality meals, which are similar to the meals one gets at catered events. On luxury cruises, the meals are cooked to order.

On the Princess line, after you've taken six cruises, you get a different color card, with which you open your cabin and pay for anything you purchase. Somehow, over the years, we've been on six Princess cruises and so have moved up to a Platinum card. You have to take sixteen cruises to move up to the next color. Princess now has replaced cards with medallions that track you through the ship and which open your cabin doors automatically. Each time you take a certain number of cruises, you get new privileges. For example, because we've taken six cruises with Princess, my wife and I will each have 240 minutes of free Internet access. Before we had free Internet, the price was sixty cents a minute, so on previous cruises, we checked our email when we were at ports in Mexico or wherever we were. The more cruises you take with Princess, the more rewards you get, so there's an inducement to keep taking cruises with Princess. On one Princess cruise we took, there was a couple who had cruised for something like 2000 days or almost five years at sea on Princess ships.

5.4 *Main Dining Room Menu on Princess Cruises*

Here is a copy of a typical dinner menu on a Princess cruise. It is similar to what you would find on cruise ships on other lines.

PRINCESS SIGNATURE PASTAS
Veal Ravioli in Creamy Porcini Mushroom Sauce
Veal Jus Fettuccine Alfredo, The All-Time Favorite from Rome
ALTERNATIVE CHOICES
Shrimp Cocktail with American Sauce
Classic Caesar Salad
Seared Salmon Cutlet with Herb Butter
Broiled Chicken Breast with Natural Juice
Grilled Beef Filet Medallions with Black Peppercorn Sauce
APPETIZERS
Crab, Scallop and Shrimp Salad featuring a Duo of Caviar, Papaya Dressing
Vegetable Paté with Baby Greens and a Lemon Grass Vinaigrette
Asparagus with Tomato and Italian Bacon with Hollandaise Sauce
SOUPS AND SALAD
Chicken and Vegetables in Fortified Broth
Roasted Tomato Cream Soup, Brioche Croutons
Ice-Cold Peach and Mango Smoothie with Dried Apricots
Belgian Endive, Boston Lettuce, and Tomato
Choice of Homemade and Low-Fat Dressings
MAIN COURSES
Pan-Fried Filet of Rockfish
Creamy Potato and Leek Casserole, Truffle Oil
Broiled Lobster Tail and King Prawns with Lemon Butter
Grilled Asparagus and Rice Pilaf
Roasted Pheasant in Pan Juices with Caramelized Shallots
Bean and Bacon Bundle, Basted Potatoes
Beef Tenderloin Wellington
Truffle-Madeira Demi-Glace, Duchesse Potatoes
Home-Style Cuisine: Red Wine Braised Beef Short Ribs Hearty Vegetables, Mashed Potatoes

Baked Crepes with Roasted Pumpkin and Walnut Mascarpone Thyme-Cream Sauce

This menu doesn't include dessert and coffee. So by the time you've had an appetizer, a salad, soup or pasta, a main course, dessert, and coffee, the hours will have slipped away—especially if you have good conversants at your dinner table and are at a large table with many people. Having good conversations at your table doesn't happen too often, unfortunately.

There is an art to carrying on a conversation and there are rules to conversing which many people on cruises do not know. One of the rules is that everyone who is involved must keep the conversation going, by answering questions in some detail (not tersely) and then by asking questions after you've answered a question. Or by bringing up a new subject for conversation. On one cruise, at one meal we were seated at a table for four with a couple that didn't know how to carry on a conversation. It was sheer torture being with them. You'd ask them a question, they would give a terse answer and then there was silence. That made us decide to ask for a table of six or more people when we dined. We chose anytime dining which meant during meal hours we could dine whenever we wanted and always had different table mates, though sometimes, by chance, we sat at a table with people we'd dined with before.

The Sailaway menu on the *Grand Princess* had a prime rib dinner and a large percentage of the people in the restaurant ordered that for their main course. We ordered the Prime Rib and it was delicious. The Sailaway menu is meant to get passengers off to a good start, as far as their dining experience is concerned. On most cruises, there is no end to prime rib dinners, surf and turf dinners, steak dinners, lobster dinners, and other treats. In the course of ten nights on the ship, you can choose from a bewildering assortment of dishes for breakfast, lunch, and dinner and if you don't like anything on the menu, on most ships you can order a steak or salmon for your main course. Sometimes nothing was interesting on the menu and at times the dishes we ordered were not very tasty. But as a rule, the food on most cruise ships varies from good to excellent. The breakfasts are always wonderful.

The amazing thing is that the mass cruise lines only spend about twenty dollars a person for food. They purchase food in such huge quantities that they end up paying relatively little for all the steaks and lobsters and other foods they purchase. The Horizon Court (cafeteria) on the *Star Princess*, one of the ships we cruised on, was open from early in the morning to late at night, so you never had to fear of going hungry. There are also hot dogs and hamburgers available most of the day, free soft ice creams, free pizzas, and one small snack bar on the ship that is open 24 hours a day. The person who served soft ice cream on one *Star Princess* cruise we took told us that on the previous cruise, there were 400

ILLUSTRATION 20 Prime rib dinner on the *Star Princess Sailaway*
PHOTO BY THE AUTHOR

ILLUSTRATION 21 Bacon and egg breakfast on the *Grand Princess*
PHOTO BY AUTHOR

young children on the boat. So he never stopped serving ice cream cones. On a cruise my son and his family took recently, over the Christmas holidays, he told me there were 700 children aboard. Some cruise lines now, such as Virgin cruises, do not allow children on board.

There is a certain kind of sameness to the food on the mass tourism lines such as Princess, Celebrity, and their competitors. I find that after a while on a cruise ship I get a bit tired of the food. It may be too much of a good thing, though there are some meals when I've found the main course to be quite different from what I expected and not particularly satisfying. But cooking for thousands of people requires some sacrifices by the cooks (of whom there are hundreds on every large ship), and as a rule, the food is good to excellent. Cruising provides passengers with a vague sense of what it is like to be rich and to have maids and cooks, all for as little as forty dollars a day (on some cruises). You share the maids (room stewards) with others, and you share your cooks with thousands of other people, but taking a cruise, gives you, for a while, a glimpse of how the rich live.

6 Cruising and the Psyche

> The North American cruise industry is a $7-billion-plus business that carries almost 5 million passengers annually and directly employs approximately 50,000 persons. It is projected that by the year 2000, as many as 8 million passengers per year will cruise. Some 200,000 travel agents sell cruises that average about $225 per person per diem. These same cruise passengers spend tens of millions of dollars on airline seats (to travel to and from their ports of embarkation), hotel rooms (for pre-and-post accommodations) and are the prime source for jewelry stores, art galleries, attractions, and gift shops in major destinations all over the world. Moreover, tens of millions are spent on board the vessels in the bars, gift shops, shore tours, casinos, and so forth. Cruising has been the fastest-growing segment of the vacation business for the last decade, beating land-based resorts, theme parks, and excursions. Studies show that whether they are first-time or frequent cruisers, the majority of travelers (63–66%) rate cruising as better than other vacation experiences. Ninety percent of them expect to take more cruises and nearly 60 percent dream of taking a cruise someday.
>
> BOB DICKINSON and ANDY VLADIMIR, *Selling the Sea: An Inside Look at the Cruise Industry*, 1997

> According to the Cruise Lines International Association (CLIA), the world's largest cruise industry trade organization, 32 million passengers are set to travel on cruise ships in 2020, up from 30 million in 2019. Since 2009, cruise ship passengers have grown from 17.8 million to 30 million, an annual growth rate of 5.4%. More significantly, cruise industry revenues have grown even faster, from approximately 15.7 billion in 2010 to an estimated 31.5 billion in 2020, a compounded rate of growth of 7.2%.
>
> Worldwide, the cruise industry generated $150 billion in direct and indirect revenues. The industry provided 1.177 million jobs and an annual payroll of $50.24 billion, an average salary of around $45,000. Contrary to the widely held belief that cruise industry jobs are low paying, the industry has among the highest average salaries in the tourism and recreation sector. There are currently 278 ocean cruise line ships operating across 55 cruise companies with another 19 ships scheduled to debut in 2020. Passenger carrying capacity is scheduled to increase by one-third between 2020 and 2025. In addition, there are now over 500 river cruise ships.
>
> https://www.forbes.com/sites/joemicallef/2020/01/20/state-of-the-cruise-industry-smooth-sailing-into-the-2020s/?sh=3f3f4ade65fa

There are many reasons why cruising is so popular and generally receives very high ratings as a form of travel from passengers. I explore some of them in this Part.

ILLUSTRATION 22
Sigmund Freud
DRAWING BY THE AUTHOR

6.1 *A Psychoanalytic Approach to Cruising*

Freudian psychoanalytic theory has some interesting things to say about cruising just as it has something to say about everything. Freud's theory postulates an unconscious in people's minds which they cannot access but which shapes much of their behavior. We can think of the mind as being like an iceberg, floating in the sea. Consciousness is the tip of the iceberg that we can see. Just below consciousness is the preconscious or subconscious, which we can dimly make out. But most of the iceberg is hidden beneath the sea, in darkness.

What is important to recognize is that material stored in the unconscious plays an important role in shaping our thinking and our behavior and much of this material is stored without our being aware of it. This material just slips into our unconscious without our consciously placing it there. And the unconscious is eighty or ninety percent of our psyches.

Gerald Zaltman, a professor at the Harvard University Business School, wrote a book, *How Customers Think: Essential Insights into the Mind of the Market*, in which he talks about the 95–5 split. He writes (2003: 50):

> Consciousness is crucial in daily life for many obvious reasons. However, an important fact and one of the key principles of this book is the 95–5 split. At least 95 percent of all cognition is below awareness in the shadows of the mind while, at most, only 5 percent occurs in higher-order consciousness. Many disciplines have confirmed this insight.

Zaltman's "shadows of the mind" is what Freud called the unconscious. It affects how consumers think and plays a larger role in our lives than we can imagine—shaping our thinking and much of our behavior.

Freud had another theory that argued that there was a constant battle in the unconscious between what he called the "id" (desire, lust), the "ego" (moderation, rationality), and the "superego" (guilt). The "id" gives us energy but lacks focus. The "superego" represents conscience and guilt, but if it is too strong, we

ILLUSTRATION 23
The iceberg model of the psyche
DRAWING BY THE AUTHOR

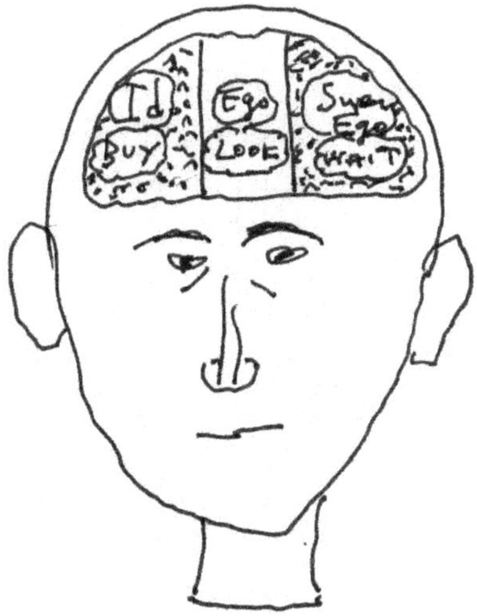

ILLUSTRATION 24
Id, ego, and superego
DRAWING BY THE AUTHOR

cannot act because we are consumed by guilt. The "ego" mediates between the "id" and the "superego." If the ego is successful we are all right but if it isn't, we have psychological problems.

Taking a cruise can be seen as essentially an "id" activity, focused upon pleasure and gratifications of various kinds. Cruising also can be seen as a regressive activity, as we return, emotionally, to when we were young and were looked after by our parents. When we were infants, we experienced non-conditional love and all our needs were taken care of. If we are dominated by our "ids" during cruising, we will eat too much, drink too much, spend too much on excursions, and gamble too much. If we are dominated by our "superegos" we won't do very much at all, feeling guilty about being on the cruise. Fortunately, for most people, our "egos" find a way to balance our "id" and "superego" elements.

In the course of conversing with our table mates, we met some passengers who had taken fifty, sixty or more cruises. That made me wonder whether there was an element of addictive and obsessive-compulsive behavior to taking so many cruises. We met one woman who was eighty-five and had been on eighty cruises. She was traveling with her son, who was sixty, who had been on fifty cruises. Ernest Jones, a Freudian psychiatrist, described obsessive-compulsive behavior as "The blind impulse to repeat earlier experiences and situations irrespective of any advantage that doing so might bring from a pleasure point of view." I wonder whether taking so many cruises brings the law of diminishing returns into operation. When you repeat something many times, the gratifications you get each time you do it diminish.

Whatever the case, I was surprised by the number of people I met on our cruises who had been on so many cruises. Cruising is a form of "soft" travel that enables people to visit all kinds of exotic countries without much stress. Ordinary travel is stressful but on cruises, as the cruise companies keep reminding everyone, you unpack once and that's it. On cruises, you can visit many different cities without having to fly between them and spend time at airports—which most travelers find to be increasingly unpleasant places.

6.2 *Come Back New*

Cruising can also be seen as a kind of "escape attempt" we make to counter the boring routines of everyday life. Princess advertises that after you've taken a cruise with them, you will "come back new."

Robert Musil, an Austrian novelist, wrote a book, *The Man Without Qualities*, in which he dealt with the problem many people face in mid-life—they feel stuck where they are. As he put it (Capricorn Books, 1965: 151):

> At this moment he wished to be a man without qualities. But this is probably not so different from what other people sometimes feel too. After all, by the time they have reached the middle of their life's journey few people remember how they have managed to arrive at themselves, at their amusements, their point of view, their wife, character, occupation, and successes, but they cannot help feeling that not much is likely to change anymore. It might even be asserted that they have been cheated, for one can nowhere discover any sufficient reason for everything's having come about as it has. It might just as well have turned out differently. The events of people's lives have, after all, only to the least degree originated in them, having generally depended on all sorts of circumstances such as the moods, the life or death of quite different people, and have, as it were, only at the given point of time come hurrying towards them—something has had its way with them like a flypaper with a fly; it has caught them fast, here catching a little hair, there hampering their movements, and has gradually enveloped them, until they lie, buried under a thick coating that has only the remotest resemblance to their original shape.

We may not think of ourselves as flies stuck on flypaper but we do face the problem, sometimes described as a "midlife crisis," of occasionally being stuck in a rut with life always, it seems, just more of the same.

Marcel Danesi discusses this topic in his *Memes and the Future of Pop Culture* (Brill, 2019: 25):

ILLUSTRATION 25 "Come back new" bag, Princess Cruises bag given to passengers
PHOTO BY THE AUTHOR

> French sociologist Emile Durkheim (1912) was the first social scientist to study the effects of ennui, claiming that such irrational boredom stemmed from a loss of moral and religious traditions. He used the term *anomie* to refer to the sense of nothingness that results from this loss. It is thus not far-fetched to envision pop culture as an antidote to boredom, with its all-encompassing thrill-seeking, comedic, sentimental, entertainment focus.

Travel is a good way to deal with boredom and cruising is a relaxed way to escape, if only for a while, from the routines of everyday life. So the Princess lines may have a point in suggesting that after taking a cruise with them, you will come back new—or, at least, renewed. Psychologists tell us that we get more pleasure from our experiences than from buying things and cruising is an easy and relaxing way to have these new experiences.

6.3 *Cruising and the Carnivalesque*

There is what can be described as the spirit of the carnival on cruise ships. There is music, there is drinking (there are more than a dozen bars on the *Grand Princess*), there are orgies of eating at the dining rooms, there is gambling, there is dancing, there are entertainments of all kinds (a piano and string trio was playing classical music on the *Star Princess* on our most recent cruise). On Holland America ships there is always a chamber music orchestra playing just about every day and sometimes more than once a day. It is this carnivalesque spirit, often helped by a bit of alcohol, which becomes infectious and explains why so many people find cruising so pleasant.

This notion of the "carnivalesque" comes to us from a Russian philosopher, Mikhail Bakhtin (1895–1975) who discussed it in his book on the novelist, Francoist Rabelais. In his book, *Rabelais and His World*, Bakhtin discusses the festive nature of medieval carnivals, which provided people with a break from their arduous everyday lives. These carnivals were attacks on authority and a means for popular renewal. As Bakhtin writes in the introduction to his book (1984: 8):

> Carnival is people's second life, organized on the basis of laughter. It is festive life. Festivity is a peculiar quality of all comic rituals and spectacles of the Middle Ages. All these forms of carnival were also linked externally to the feasts of the Church. (One carnival did not coincide with any commemoration of sacred history or of a saint but marked the last day before Lent, and for this reason, was called *Mardi Gras*).... Even more

ILLUSTRATION 26
Mikhail Bakhtin
DRAWING BY THE AUTHOR

> significant is the genetic link of these carnivals with ancient pagan festivities, agrarian in nature, which included the comic element in their rituals. The feast (every feast) is an important primary form of human culture. It cannot be explained merely by the practical conditions of the community's work, and it would be even more superficial to attribute it to the physiological demand for periodic rest. The feast had always an essential, meaningful philosophical content.

What Bakhtin suggests is that there is an important philosophical meaning to carnivals and the feasts and other activities that are tied to them. He also adds that elements of the carnival exist today. Some historians have suggested that it was the feasts, and all the work needed to put them on, which led to the development of civilizations and modern societies.

We find these elements in many festive celebrations, such as the *Mardi Gras* celebrated in many cities and, as I've suggested in this book, as a dominating motif, in cruises. On cruise ships, the carnivalesque spirit is created by the Carnival Corporation and the staff of the ship. It is also a function of the nature of cruising—as a resort experience. By a curious coincidence, the company that owns the Princess Line is named Carnival. The Carnival Corporation also owns the Carnival line, the Holland American line, the Seabourn luxury line, the Cunard line, the Aida line, the Costa line, and the P&O line. Carnival dominates the cruise industry.

6.4 *Cruising and Addiction*

This article, taken from one of the many Internet cruise sites, deals with a topic that I have speculated about, the addictive nature of cruising. Why some people take so many cruises is a topic that I became interested in when talking to passengers on the *Grand Princess* many years ago.

> 1. HOW ADDICTING CRUISES ARE—Cruises are the most addicting type of vacation that you can take. From the ocean breeze, to arriving in a new port each morning, and the fun activities offered on board, one cruise is all it takes to start the addiction. Do you work to support your cruise addiction?
>
> 2. BOOKING A CABIN WITHOUT WINDOWS IS NORMAL—A room without any windows? Yes, inside staterooms are not only available on cruises, but are preferred by many cruisers. The value that they offer allows them to cruise more often. And besides that, how much time do you spend in your cabin on a cruise anyway?

3. DOING NOTHING ON A SEA DAY IS ACTUALLY DOING SOMETHING—Ever hear, "I would take a cruise, but I don't want to be stuck on a ship"? Sea days quickly become your favorite days of a cruise.

4. WAVING NAPKINS AT DINNER—As soon as Funiculi Funicula starts playing at dinner, you immediately start waving a white napkin over your head. Waving napkins is also encouraged during the Baked Alaska parade.

5. MENUS WITHOUT PRICES—Menus with prices? No thanks, we'll leave those for land based restaurants.

6. HAVING MULTIPLE CRUISES BOOKED AT ONCE IS NORMAL—Land based vacations are normally booked one at a time. However, a cruiser can have several cruises currently booked. It is also common for some to disembark one ship and immediately embark on different cruise ship or the same ship on the same day.

7. FINDING EMPTY LOUNGERS BY THE POOL ON A SEA DAY FEELS LIKE HITTING THE LOTTERY—This is unfortunately true due to the rampant saving of chairs on the lido deck.

http://cruisefever.net/0802-16-things-only-people-obsessed-with-cruising-will-understand

This material is meant to support going on cruises and counters the commonly heard refrain that "Cruises are boring." I dealt with this matter earlier in the book. I don't know whether the people who take fifty or a hundred cruises are cruise addicts or obsessive-compulsive cruise takers, but something is interesting about the fact that some people take so many cruises, at considerable expense in money and time. I've met some people on our cruises who told me that they take five or six cruises a year and have done so for many years.

6.5 *Cruising and Quantum Experiences*

Many tourism scholars have found that people who travel hope to have is what is called a "peak experience." An example would be a chance encounter with someone at a site or a fantastic meal in a local restaurant that generates happiness and a sense of fulfillment. These are among the most memorable people take from their trips. But there is another experience that people seek, often without recognizing what they are searching for, what is called a "quantum change," which is a more profound and long-lasting matter than a peak experience. A quantum is an experience that leads to a profound change in a person's sense of themselves, their values, their beliefs, and their perspective on life. An article by Daniela Hernandez, "A Trip to Antarctica Changed My Life," in *The Wall Street Journal* (June 6, 2020) described how her trip to Antarctica radically changed her life. She flew to Antarctica, spent two weeks there, and experienced a quantum change. She returned a different person. She made many important changes in her values, beliefs, and other aspects of her life after her trip. I would guess that many people who take cruises are, without recognizing it, really in search of that kind of experience.

When Princess gives people bags saying "Come Back New," this slogan means refreshed, with one's batteries recharged, but for people who have quantum experiences, they come back *really* new.

7 Princess Cruises: A Case Study

> There are a variety of tourist glances, the capturing of sights in passing from a railway carriage, through the car windscreen, the steamship porthole, or the camcorder viewfinder (see Larsen, 2001). As Schivelbusch argues: 'the traveller sees ... through the apparatus which moves him through the world.
>
> http://www.lancs.ac.uk/fass/sociology/papers/urry-globalising-the-tourist-gaze.pdf

> Name an activity that really should take place on solid ground. Roller coaster would seem like a good answer. Until now.... When the *Mardi Gras* sails late in 2020, riders will zip along at up to 37 miles per hour above the top deck, with turns and drops that make it appear like you are plugging into the ocean.... Carnival's cruising coaster is the latest move in an activity arms race aboard cruise ships. These massive vessels have become floating amusement parks offering 10 story slides go-kart tracks, surfing, waterslides, ziplines and simulated sky diving. To appeal to families, especially multigenerational families, ships have to offer a variety of experiences.
>
> SCOTT MCCARTNEY, "The Roller Coaster That Floats." *The Wall Street Journal*. January 9, 2020

Twenty-five new cruise ships will enter service in 2020, representing over 38,000 new berths entering the global cruise market, according to the *2020 Cruise Industry News Annual Report*. The smallest of the new vessels is the 120-guest Coral Geographer, built for expedition itineraries in Australia, while the largest vessel is the 5,200-guest Carnival Mardi Gras,

ILLUSTRATION 27 The *Grand Princess*
PHOTO BY THE AUTHOR

which will run on clean-burning LNG, sailing Caribbean cruises from Port Canaveral.

Cruise Industry News, Jan. 4, 2020

7.1 *A Cruise Critic Review of the Grand Princess*

Cruise Critic is an Internet site that reviews cruise ships. It carries "expert" reviews, written by a professional journalist, and also reviews by people who have taken cruises on each ship. What follows is a Cruise Critics overview of the ship written by an expert reviewer.

Updated by Carolyn Spencer Brown, Editor in Chief

Editor Rating 4.0

When *Grand Princess* was being built back in 1998, it was described by the company as "the biggest, the fastest, the most elaborate, the most technologically advanced ... the grandest ship on the ocean." My, how times have changed. *Grand Princess* did indeed spend a year as the biggest, fastest, and most elaborate ship at sea, but it is now dwarfed by many others that are bigger, faster, newer, and more technologically sophisticated. And yet, more than a decade later, the basic ship design that anchors Princess' most popular class (which includes *Caribbean Princess, Star Princess, Golden Princess, Crown Princess, Ruby Princess,* and *Emerald Princess*) still holds its own thanks to a massive overhaul in May 2011. The multi-million dollar refurbishment of *Grand Princess* included a complete

> transformation of the ship's atrium into the much jazzier Piazza (a newfangled addition introduced by the fleet's most recent ships that combine theater in the round with food and drink outlets). It finally got its own version of the Sanctuary, a popular adult-only sun deck that debuted on Crown Princess in 2006. Other additions included new restaurants and lounges ranging from Alfredo's Pizzeria to the Tea Leaves tea salon, the Crown Grill steakhouse, and the snazzy One5 nightclub. A seven-night cruise from Southampton gave me ample opportunity to check out the new and improved *Grand Princess*. There was barely a nook or cranny that didn't get some sort of special overhaul—with two exceptions. One of the biggest complaints we heard on board was that the central bank of elevators was old, creaky, and unreliable. (Two were out of service during our cruise, which didn't help matters.) Because the original design of *Grand Princess* did not include a central stairwell servicing all decks, you've either got to walk forward or aft from the middle of the ship or wait, interminably, for a lift. The second area that did not get much love was cabins. The public areas received the lion's share of attention (the massive refurbishment also addressed mechanical systems and such), and there has been much chat on our Princess' forums about how dated the cabins continue to look.
>
> **Cruiser Rating**
> Excellent

The passenger reviews ranged from people who gave it one star (out of five) and hated the cruise to others who gave it five stars and loved it. People who write reviews of ships on Internet sites tend to be either disgruntled former passengers who feel very negative about their cruise or those who feel very positive about it. According to Cruise Critic, 74% of the passengers on its cruises "loved it."

7.2 *The Grand Princess Alaska Cruise*

My wife and I took a cruise from San Francisco to Alaska and back on the *Grand Princess*, from May 26 to June 5, 2019. We had checked the weather before we left and were expecting rain in many of the ports, but it didn't rain at all during the voyage. I should add that this was our third trip with Princess to Alaska.

The one before this trip was fourteen years ago on the first *Regal Princess*, which no longer is part of the Princess fleet. Travel, we must remember is a marker of status, so the passengers on the ship tended to be from a certain socio-economic class, which I would guess would be upper-lower class to lower-upper class. On the cruise, we dined, at various times, with a provost of a university, a person with a Ph.D. in psychology, and another person with a Ph.D. in another subject. They were the exceptions, not the rule. Most of the passengers were office managers, blue-collar workers, and people with similar occupations. The passengers didn't "look" like members of the upper classes. As ocean cruises go, Princess and the other lines like Princess are relatively inexpensive.

Princess tends to serve Prime Rib during the breakaway dinner and everyone at our table of six, it turned out, had the Prime Rib dinner. I had a shrimp cocktail, gravlax, a salad, and the Prime Rib, medium-rare. Plus dessert. The Prime Rib is a huge and gorgeous slab of meat, served with baked potato and corn. You do not go away hungry from that meal. We were at a table with two widows traveling together and two single women. I was the only male at the dinner table. It turned out that we had a very good conversation at dinner. Everyone at the table was a good conversationalist and my wife and I enjoyed their company.

Unfortunately, I cannot say that for all of the other conversations we had with other tablemates. It was cold the next day. We had breakfast, also at a table for six, with a couple from Georgia and a woman from Australia who was a non-stop talker. One of many non-stop talkers we encountered during the trip. We went to a fruit carving exhibition in the Atrium, had lunch, and took a nap. I don't know whether it was the sea air or the excitement of the trip, but during the trip, we often had naps—enjoying what I would describe as an eat/sleep/talk cruise. At dinner, we sat with a woman who kept repeating, many times, "We're lower class," which I found curious. Was she telling the truth or trying to be amusing? Next to her was someone with an academic background, with whom my wife, who taught philosophy, and I had a long conversation about philosophy. So we had a strange collection of people at that meal. After dinner, we always went to the theater, where there were musical shows, with wonderful singers and dancers, comedians, jugglers, and other kinds of entertainment. We went to the theater every night except one and that was when the comedian Princess brought on board was giving a second show which didn't interest us. I thought he was terrible.

The first formal night dinner was mediocre, at best. I would say that the food on the ship varied from excellent, most of the time, to just tolerable, some of the time. Fine dining is, of course, one of the most important parts of cruising for most people. The larger your table, the more time you spend in the dining room since the waiters have to serve so many passengers.

My wife and I walked around the cities we docked at Ketchikan, Juneau, Skagway, and Victoria. The weather was often a bit chilly in the morning but generally, quite pleasant by noon and the rain we expected never came. The Alaskan cities are full of diamond shops and souvenir shops. The areas we walked around were all old and looked like they were built in the early 20th century. I would describe them as "funky." In some of the Alaskan cities we visited, there were six ships, from different lines, at the port. The shops often had blackboards listing the names of the ships or the cruise lines with ships docked at the port. Walking around the cities in Alaska was a kind of momentary return to the past and an investigation of people living a much different lifestyle from my wife and me and probably most of the passengers on the ship. These cities are very touristy but they are also charming and have an aesthetic quality to them that we found very pleasing. And in some cities, you had gigantic snow-capped mountains in the background.

The Alaska cruise is very popular and the scenery, especially when you get to see the glaciers, is quite spectacular. There is considerable competition among the lines for passengers, many of whom board in Seattle or a Canadian port. We took the Princess cruise because we live only a twenty-five-minute drive from the port in San Francisco. Many of the passengers on the *Grand Princess* told us that they took the cruise because they live near San Francisco and didn't want to fly to take a cruise, to the extent that they could avoid doing so. The highlight of the cruise for us was the glacier. We had breakfast at a table near a window on the ship and facing us was the glacier—a gigantic body of ice that was awe-inspiring.

When the ship docks in Victoria it is a pleasant walk from the pier to the center of the city, where there is a wonderful hotel, the Empress Hotel, which serves an elegant tea service in the afternoon. It is located in an area where there are various government buildings and many other hotels, but none with the glamor or panache of the Empress Hotel. It took us 10,000 steps to walk from the ship to the hotel and back, but it was a beautiful sunny day and the walk was quite exhilarating.

ILLUSTRATION 28
Glacier, Alaska
Photo by the author

ILLUSTRATION 29 The Fairmont Empress Hotel, Victoria, Canada
PHOTO BY THE AUTHOR

7.3 *Two Splurges*

For some reason, I was given $400 to spend on the cruise by the Princess line. I got $100 for having been in the US Army and another $300 as an incentive to take other cruises. So we decided to dine in the two specialty restaurants on the ship: the Crown Grill and Sabatini's. Our dinner at the Crown Grill, a steak house, was good but not great, by any means. My wife had lamb chops and I had a ribeye steak that was too fatty and didn't have much taste. I don't think it was worth the $29 each we paid to eat there.

Our other dinner at Sabatini's was superb; it was much better than our dinner at the Crown Grill. Sabatini gives you a free starter course and then you can order various dishes to suit your fancy. Everything was tasty and the dinner lasted two hours. The restaurant has a new and more Italian menu on some ships and they had the new menu on the *Grand Princess*. This description of the new menu is taken from the Princess website.

> **New Menu**
>
> Aboard Caribbean Princess, Grand Princess, Regal Princess, Coral Princess and Star Princess, cruisers will enjoy a new Sabatini's menu, designed

> by Italian chef Angelo Auriana. Like most Italian restaurants, the meal begins with bread. This time it's a rustic whole-wheat loaf, Tuscan zonzelle with roasted tomato and bell pepper passata, and prosciutto crudo served with olive oil and balsamic vinegar. Diners then choose an antipasto, soup or salad, pasta, and main course. Antipasti treats include fried zucchini and yellow squash, smoked scamorza, and housemade crostini as well as chickpea fritters with squid, prawns, eggplant, carrots, green beans and a lemon garlic aioli. The next course features a celery, mushroom and toasted hazelnut fondue or an escarole salad with celery, pear and orange zest. The pasta course offers a handful of choices, like risotto with beets, gorgonzola and pistachios or buckwheat pappardelle with braised beef cheek. The entrees also include items such as lake trout rolls with rosemary polenta, porchetta, breaded veal vallet, a vegetarian polenta with mushrooms, mozzarella and truffle, and more. Desserts change throughout the voyage so request the menu to see what's on offer on the evening you're dining there.

This dinner on the *Grand Princess* cost $25 per person and was the best Italian meal I've had in years. It was a kind of theater, in which my wife and I were the main actors. Everything on the menus was delicious and after dining at Sabatini's, we felt we'd had a remarkable experience. We had eaten at Sabatini's years before and found it excellent, but its new menu was a treat. I taught at the University of Milan in 1963 for a year and this dinner reminded me of meals I'd had while in Italy.

After dinner, we went to see a really fine show, "Born to Dance," which featured the ship's dancers. It was the best of the three song-and-dance production shows we saw while on the cruise. Generally speaking, we found the entertainment on the ship to be first-rate.

I particularly liked a Hungarian string trio that often entertained in the Atrium. The trio consisted of a violinist, a bassist, and a pianist and they had an immense repertoire. The violinist asked people for songs they wanted and the trio played them. When he was playing one song, a woman who was sitting in the Atrium got so excited she stood up and sang the words to the song. The violinist smiled and was amused and after she sang, people in the audience applauded. Every night there was some kind of entertainment in the theater and every night, many of the passengers who dined early made the trek to the other end of the ship to watch the shows. There were singers, comedians, and jugglers who entertained us. The juggler was really a comedian who did a bit of juggling. He was a very polished entertainer and "played" with the audience.

Some of our tablemates told us that they had been on many different lines but preferred Princess since they thought it offered the most value for the dollar. And, like most lines, Princess offered incentives to get passengers to keep coming back. I was surprised by the number of people we met who had accumulated enough days to become elite passengers and have their laundry done for free. I would describe the Princess line as a quality line with good food (most of the time), generally excellent entertainment, that offers, as one passenger put it, "a big bang for the buck." The *Grand Princess* is, as I've already mentioned, an old ship and shows its age here and there, but it is in pretty good condition. We decided that we had a good time on the cruise and wondered how it would compare with our forthcoming cruises with Celebrity and MSC.

Our Alaska cruise was not expensive, as cruises go, and we enjoyed the scenery and the small, funky Alaskan towns. Many people who are committed Princess cruise-takers keep telling me the same thing. "Princess has gone downhill since it was purchased by Carnival," but they still sail with Princess because they believe it is the best choice. And, as I pointed out earlier after you sail with Princess you get certain benefits, which keep adding up, so in many cases, people sail with Princess because of the benefits they get: free Wi-Fi, free laundry, money to spend on restaurants, and excursions, and so on.

The Carnival Line now has a ship, the *Carnival Miracle*, which sails from San Francisco and offers cruises that are slightly less expensive than the Princess cruises. And Princess has moved the *Grand Princess* to another port and substituted a different ship, the *Ruby Princess that* sails from San Francisco. Both lines are owned by the Carnival Corporation, which must feel there are enough people in the area who would like to cruise to make it reasonable for them to have two cruise ships in San Francisco.

8 Coda

> Cruise ships, which now cost upward of $1 billion to build, present a significant challenge. Repeat customers and social media reviews are crucial to the business. So comfort for the masses actually matters and it drives ship design. Carnival, the largest cruise company, is building its largest ship, a 180,000 ton floating town three football fields long with 18 decks and even the first roller coaster at sea. The *Mardi Gras* which will sail in late 2020, will carry a maximum of 6,641 passengers in 2,641 cabins.
>
> SCOTT MCCARTNEY, "Making 180,000 Tons Feel Cozy at Sea." *Wall Street Journal.* December 19, 2019

> Analysis of tourism and tourist culture has tended to work through a purely visual framework, responding to the perpetual visualization in tourism-related discourses. Indeed, Urry (1990) argues that tourism is all about gazing upon particular scenes that are different from those encountered in everyday life. Furthermore, this view of tourism has led Urry to the conclusion that since postmodernity is marked by "the proliferation of images and symbols" ... tourism is coming home, since it is now possible "to *see* many of the typical objects of the tourist gaze ... in one's own living room."
>
> HAZEL TUCKER, "The Ideal Village: Interactions through Tourism in Central Anatolia." In Simone Abram, et al, *Tourists and Tourism: Identifying with People and Places* (1997: 107)

> Carnival is the people's second life, organized on the basis of laughter. It is a festive life. Festivity is a peculiar. quality of all comic rituals and spectacles of the Middle Ages.... The feast (every feast) is an important primary form of human culture. It cannot be explained merely by the practical conditions of the community's work, and it would be even more superficial to attribute it to the physiological demand for periodic rest.
>
> MIKHAIL BAKHTIN, *Rabelais and His World*

The photo of the balloon "orgy" at the end of a Princess cruise captures the sense of joy and pleasure people in the theater felt after seeing a show and being allowed to play with balloons. Balloons were flying all over the place and people were smacking them in the air with great relish. It is an image, I believe, that shows what Bakhtin's carnivalization theory is all about.

In this book I've done two things: First, I've discussed why people travel and become tourists. You might say you have to travel—that is, go somewhere—to become a tourist. And second, I've written about the pleasures of cruising and offered some semiotic, sociological, and psychoanalytically informed ideas about why people take cruises and why some people take so many cruises.

Tourism, from a semiotic and somewhat reductionist perspective, involves a search for signs of interest in different places—destinations like Bali or Angkor Wat, buildings like the Eiffel Tower, cities such as Barcelona or Tokyo, wonderful hotels, restaurants, beaches, and deserts; wherever tourists can have the kind of experiences they seek. The sign becomes of central concern for a semiotic understanding of tourism and one can argue that the consumption of signs, in one form or another—beautiful vistas, people in native garb, foreign cuisines, and so on—all can be seen as a succession of photo opportunities. Photography, we can say, is a means of taking the tourist gaze and making it a

ILLUSTRATION 30 Balloon Orgy after the last show on the *Grand Princess*
PHOTO BY THE AUTHOR

permanent part of the tourist's recorded experience. It can be argued that the meaning of travel becomes transformed by photographic possibilities and in some cases, the places tourists visit becomes secondary, in their minds, to their photographic potential.

In a sense, tourists are semioticians who "consume" signs of historical significance and important destinations and take possession of them when they photograph or make videos of the places they have visited—manifestations of what John Urry calls "the tourist gaze." As tourists move about the world they are, in a sense, writing and photographing important events in their lives. Many tourists take photographs of themselves in front of the places they visit, showing an identification with them and also proving proof that they have been to these sites.

On the psychological level, tourism represents a kind of escape attempt from the routines of everyday life that enriches tourists and enables them to return enriched with an enhanced view of life's possibilities. Tourism also gives travelers a positive identity—world traveler, to add to the other identities they have. I dealt with the gratifications that tourism provides earlier in the book. Tourists relish the experiences they have had and anticipate the ones they expect to have, especially "peak" experiences they occasionally have in the course of their travels. Travel is often hard work but the psychological payoffs are so powerful that people are willing to put up with long flights and discomforts of one sort or another to obtain the gratifications they seek.

Tourists generally do a good deal of planning before they decide on where they want to go, what they want to see, and how they want to see it—on a tour or traveling on their own. They have been influenced by what MacCannell calls "markers" such as advertisements, guidebooks, videos, Internet travel sites like TripAdvisor and articles and books by travel writers. Many tourists keep journals and some write long and elaborate travel reports on sites such as TripAdvisor forums that have information about sites to visit, the best hotels and restaurants, getting from airports to hotels or rented apartments, and so on. Some forums have elaborate and detailed trip reports in which visitors list everything they have done during a visit to some city, including all the restaurants they dined at, sites of interest they visited, plus tips on where to go and what to avoid. So the trip, itself, is only the tip of the travel iceberg, and, from the statistics I offer below, we can see that tourism has moved from being on the periphery of the world's economy to a position of centrality. Tourism plays an all-important role in the world's economy and is a driver of economic growth in many countries.

According to the United Nations World Tourism Organization (UNWTO), in 2018 there were 1.491 billion international trips. And there were countless trips within countries. In 2019, according to Statista.com, the tourism industry was valued at 9.25 trillion dollars. Let me offer some data, on where tourists go and how much money they spent on their travels (before the pandemic, that is).

8.1 *The UNWTO on Tourism*

The top five travel destinations were, according to the UNWTO Statistics of Tourism:

Country	Number of tourists (in millions)
89	France
83	Spain
80	USA
63	China
62	Italy

The earnings from the top five tourist destinations were:

Country	Earnings in dollars (billions)
214	USA
74	Spain
67	France

63	Thailand
52	UK

The countries whose citizens spent the most money on tourism are:

Expenditures	**Country**
277 billion	China
144 billion	USA
94 billion	Germany
76 billion	UK
48 billion	France

Chinese tourists spend almost twice as much money when traveling as do people from the United States, with Germany, the UK, and France being next in line in terms of national expenditures. It is obvious that a large percentage of travelers and tourists come from Western Europe, the United States, and China.

The cruise industry, with almost 28 million passengers and hundreds of thousands of people who work on the ships, plays an important role in world tourism and until the coronavirus hit, was expanding rapidly and enjoying great success. It faced very rough seas during the pandemic but bookings for 2021 and 2022 are high. A report from the *Cruise Industry News* on July 1, 2021, suggests that cruising is back and soon will be close to pre-pandemic norms:

> July will see the quick restart of the cruise industry continue with 141 cruise ships from 50 different operators set to sail with paying passengers aboard, according to the July 2021 edition of the *Cruise Ships in Service Report by Cruise Industry News*. The 141 ships represent over 180,000 berths at full occupancy.

So, for the cruise industry, it looks like it will be smooth sailing again.

8.2 *How I Wrote This Book*

Like many writers, I keep a journal in which, among other things, I plan my books. Thus, for *Smooth Sailing*, I had some pages in my journal on which I played around with ideas, listed topics I might discuss in it, and so on. Some pages in my journals are solid blocks of writing but on other pages, I brainstorm, make charts, make lists, and on some pages, make drawings. While I was writing the book, at the same time, in my journal, I was writing about writing the book. Below, I offer two pages from the journal I kept while writing the book to give you an idea of what the process of wiring a book is like.

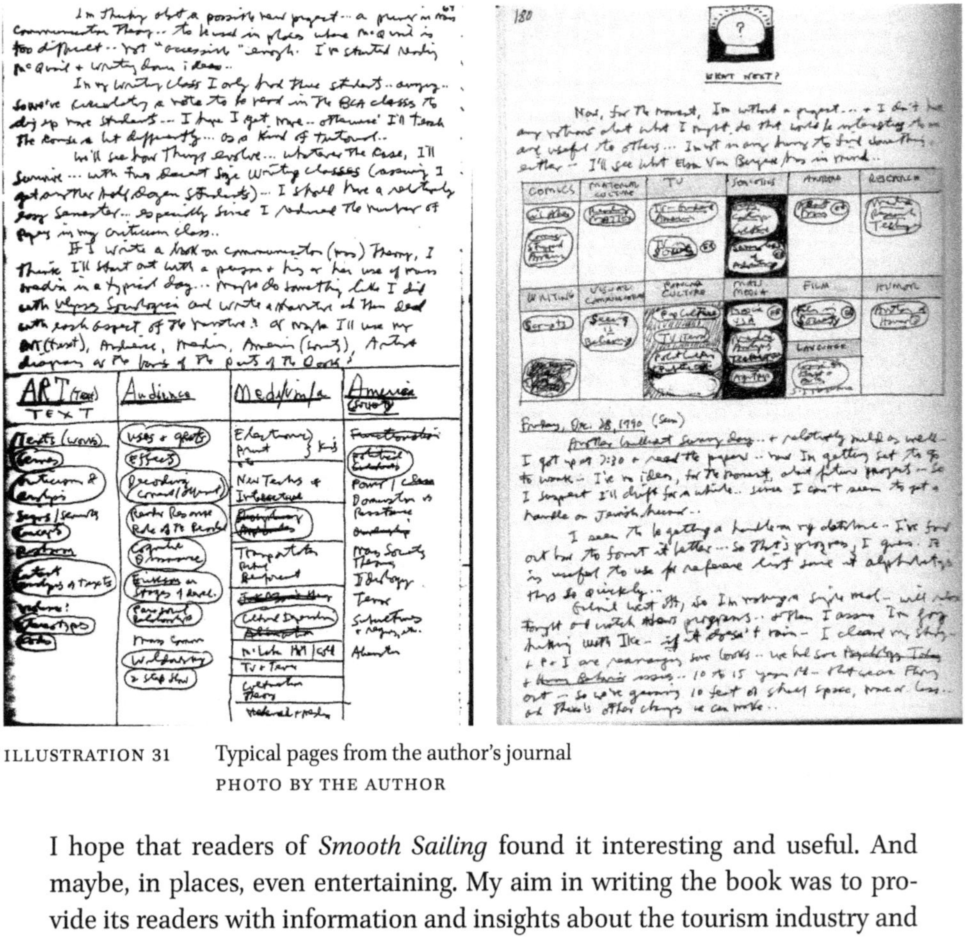

ILLUSTRATION 31 Typical pages from the author's journal
PHOTO BY THE AUTHOR

I hope that readers of *Smooth Sailing* found it interesting and useful. And maybe, in places, even entertaining. My aim in writing the book was to provide its readers with information and insights about the tourism industry and ocean cruising which will help them decide, should they wish to travel, the best way for them to do so.

References

Bakhtin, Mikhail. (1993) *Rabelais and His World.* Bloomington: Indiana University Press.

Barthes, Roland. (1972). *Mythologies.* (A. Lavers, transl.) New York: Hill & Wang.

Barthes, Roland (1982). *Empire of Signs.* New York: Hill & Wang.

Baudrillard, Jean. (1998). *The Consumer Society: Myths & Structures.* Thousand Oaks, CA: Sage.

Brown, Carolyn Spencer. Review of the *Grand Princess.* CruiseCritic.Com.

Bufford, Bill. "Editorial." *Granta: Travel Writing*. 1984. https://www.amazon.com/Granta-10-Writing-Bill-Buford/dp/0140070524.

Boorstin, Daniel. (1961). The *Image: A Guide to Pseudo-Events in America*. New York: Vintage.

Camus, Albert. (1963). *Notebooks, 1935–1942*. New York: Alfred Knopf.

Cohen, Erik. "The Changing Faces of Contemporary Tourism." *Society*. Vol. 45, No. 4. July/August 2008.

Cohen, Erik. "Toward a Sociology of International Tourism," in Kotler, Phillip, John Bowen, and James Markens. *Marketing for Hospitality*. 2nd edition. Upper Saddle River, NJ: 1999.

Cruise Industry News. July 1, 2021.

"Cruising and Addiction." http://cruisefever.net/0802-16-things-only-people-obsessed-with-cruising-will-understand.

Dickinson, Bob, and Andy Vladimir. (1997). *Selling the Sea: An Inside Look at the Cruise Industry*. New York: John Wiley.

Douglas, Mary. "In Defence of Shopping" (in Pasi Falk and Colin Campbell, editors, *The Shopping Experience*, 1997). Thousand Oaks, CA: Sage.

Eliade, Mircea. (1961). *The Sacred and the Profane: The Nature of Religion*. New York: Harper & Row.

Fussell, Paul. (1987). *The Norton Book of Travel*. New York: W.W. Norton. https://www.nytimes.com/2019/07/26/travel/bahamas-budget-cruise.html.

Glusac, Elaine. "A Three Day Cruise to the Bahamas for $210." *The New York Times*. July 28, 2019. https://www.nytimes.com/2019/07/26/travel/bahamas-budget-cruise.html.

Hagerty, James A. "Entrepreneur Sold Cruises to Middle-Class Americans." Obituaries. The *Wall Street Journal*. Nov. 17–18, 2019.

Hernandez, Daniela. "A Trip to Antarctica Changed My Life." June 6, 2020. *The Wall Street Journal*.

Jemsel, Samuel. (1930). In Elkan Nathan Adler's *Jewish Travelers in The Middle Ages*. New York: Dover.

Leed, Eric J. (1991). *The Mind of the Traveler: From Gilgamesh to Global Tourism*. New York: Basic Books.

Lėvi-Strauss, Claude. *Tristes Tropiques: an Anthropological Study of Primitive Societies in Brazil*. New York: Atheneum. 1970.

MacCannell, Dean (1976). *The Tourist: A New Theory of the Leisure Class*. New York: Schocken Books.

McCartney, Scott. "Making 180,000 tons Feel Cozy at Sea." *The Wall Street Journal*. Dec. 19, 2019. Page A11.

Musil, Robert. (1965). The *Man Without Qualities*. Vol. 1. New York: Capricorn.

New Menu. Princess.com.

"Overtourism Roundtable," *Travel Weekly*, 7/10, 2018er.

Personality. *Wikipedia.*

Plog, Stanley. "Why Destinations Rise and Fall in Popularity." https://journals.sagepub.com/doi/abs/10.1177/001088047401400040.

Sampson, Heather. "What does America have against vacations?" *The Washington Post.* April 28, 2019.

Saussure, Ferdinand de. (1966) *Course in General Linguistics.* (W. Baskin, Trans.) New York: McGraw-Hill.

Shorer, Mark. https://www.coursehero.com/file/p7tmt80/Mark-Shorer-Myth-is-fundamental-the-dramatic....

Tucker, Hazel. "The Ideal Village: Interactions through Tourism in Central Anatolia" (in Simone Abram, et al, *Tourists and Tourism: Identifying with People and Places.* 1997: 107).

UNESCO Convention for the Safeguarding of the Intangible Cultural Heritage.

Urry, John. "Globalizing the Tourist Gaze." http://www.lancs.ac.uk/fass/sociology/papers/urry-globalising-the-tourist-gaze.pdf.

Wildavsky, "A Cultural Theory of Preference Formation" (in A. Berger, Ed. *Political Culture and Public Opinion.* 1989) New Brunswick, NJ: Transaction Books.

World Tourism Organization (http://www.world-tourism.org/statistics/tsa_project/TSA).

Zaltman, Gerald. (2003) *How Customers Think: Essential Insights into the Mind of the Market.* Cambridge, MA: Harvard University Business School Press.

www.ingramcontent.com/pod-product-compliance
Lightning Source LLC
Chambersburg PA
CBHW071356230926
51846CB00002B/574
* 9 7 8 9 0 0 4 5 0 6 2 7 5 *